The Analysis of Means

ASA-SIAM Series on
Statistics and Applied Probability

ASA
SIAM

The ASA-SIAM Series on Statistics and Applied Probability is published jointly by the American Statistical Association and the Society for Industrial and Applied Mathematics. The series consists of a broad spectrum of books on topics in statistics and applied probability. The purpose of the series is to provide inexpensive, quality publications of interest to the intersecting membership of the two societies.

Editorial Board

Nelson, P. R., Wludyka, P. S., and Copeland, K. A. F., *The Analysis of Means: A Graphical Method for Comparing Means, Rates, and Proportions*

Burdick, R. K., Borror, C. M., and Montgomery, D. C., *Design and Analysis of Gauge R&R Studies: Making Decisions with Confidence Intervals in Random and Mixed ANOVA Models*

Albert, J., Bennett, J., and Cochran, J. J., eds., *Anthology of Statistics in Sports*

Smith, W. F., *Experimental Design for Formulation*

Baglivo, J. A., *Mathematica Laboratories for Mathematical Statistics: Emphasizing Simulation and Computer Intensive Methods*

Lee, H. K. H., *Bayesian Nonparametrics via Neural Networks*

O'Gorman, T. W., *Applied Adaptive Statistical Methods: Tests of Significance and Confidence Intervals*

Ross, T. J., Booker, J. M., and Parkinson, W. J., eds., *Fuzzy Logic and Probability Applications: Bridging the Gap*

Nelson, W. B., *Recurrent Events Data Analysis for Product Repairs, Disease Recurrences, and Other Applications*

Mason, R. L. and Young, J. C., *Multivariate Statistical Process Control with Industrial Applications*

Smith, P. L., *A Primer for Sampling Solids, Liquids, and Gases: Based on the Seven Sampling Errors of Pierre Gy*

Meyer, M. A. and Booker, J. M., *Eliciting and Analyzing Expert Judgment: A Practical Guide*

Latouche, G. and Ramaswami, V., *Introduction to Matrix Analytic Methods in Stochastic Modeling*

Peck, R., Haugh, L., and Goodman, A., *Statistical Case Studies: A Collaboration Between Academe and Industry, Student Edition*

Peck, R., Haugh, L., and Goodman, A., *Statistical Case Studies: A Collaboration Between Academe and Industry*

Barlow, R., *Engineering Reliability*

Czitrom, V. and Spagon, P. D., *Statistical Case Studies for Industrial Process Improvement*

The Analysis of Means

A Graphical Method for Comparing Means, Rates, and Proportions

Peter R. Nelson
Clemson University
Clemson, South Carolina

Peter S. Wludyka
University of North Florida
Jacksonville, Florida

Karen A. F. Copeland
Boulder Statistics
Boulder, Colorado

Society for Industrial and Applied Mathematics
Philadelphia, Pennsylvania

American Statistical Association
Alexandria, Virginia

The correct bibliographic citation for this book is as follows: Nelson, Peter R., Peter S. Wludyka, and Karen A. F. Copeland, *The Analysis of Means: A Graphical Method for Comparing Means, Rates, and Proportions*, ASA-SIAM Series on Statistics and Applied Probability, SIAM, Philadelphia, ASA, Alexandria, VA, 2005.

Library of Congress Cataloging-in-Publication Data

Nelson, Peter R.
    The analysis of means : a graphical method for comparing means, rates, and proportions / Peter R. Nelson, Peter S. Wludyka, Karen A. F. Copeland.
        p. cm. — (ASA-SIAM series on statistics and applied probability)
    Includes bibliographical references and index.
    ISBN 0-89871-592-X (pbk.)
      1. Analysis of means. I. Wludyka, Peter S. II. Copeland, Karen A. F. III. Title. IV.
    Series.

QA279.N45 2005
519.5—dc22   2005049966

Excel is a trademark of Microsoft Corporation in the United States and/or other countries.

SAS and SAS/QC are registered trademarks of SAS Institute Inc.

MINITAB is a registered trademark of Minitab Inc.

Figures A.1–A.8 in Appendix A are reprinted with permission of the publisher from P. R. Nelson, "Power Curves for the Analysis of Means," *Technometrics*, Vol. 27, No. 1, February 1985, pp. 65–73. © 1985 by the American Statistical Association. All rights reserved.

Figures A.9–A.16 in Appendix A are reprinted with permission of the publisher from P. S. Wludyka, P. R. Nelson, and P. R. Silva, "Power Curves for the Analysis of Means for Variances," *Journal of Quality Technology*, Vol. 33, 2001, pp. 60–65. © 2001 American Society for Quality.

Figures A.17–A.24 in Appendix A are reprinted from E. J. Dudewicz and P. R. Nelson, "Heteroscedastic Analysis of Means (HANOM)," *American Journal of Mathematical and Management Sciences*, Vol. 23, 2003, pp. 143–181. Copyright © 2003 by the American Sciences Press, Inc., 20 Cross Road, Syracuse, NY 13224. Reprinted by permission.

Table B.1 in Appendix B is reprinted with permission of the publisher from P. R. Nelson, "Additional Uses for the Analysis of Means and Extended Tables of Critical Values," *Technometrics*, Vol. 35, No. 1, February 1993, pp. 61–71. © 1993 by the American Statistical Association. All rights reserved.

Table B.2 in Appendix B is reprinted with permission of the publisher from P. R. Nelson, "A Comparison of Sample Sizes for the Analysis of Means and the Analysis of Variance," *Journal of Quality Technology*, Vol. 15, 1983, pp. 33–39. © 1983 American Society for Quality.

Table B.3 in Appendix B is reprinted with permission of the publisher from P. R. Nelson, "Multiple Comparisons of Means Using Simultaneous Confidence Intervals," *Journal of Quality Technology*, Vol. 21, 1989, pp. 232–341. © 1989 American Society for Quality.

Table B.4 in Appendix B is reprinted with permission of the publisher from P. S. Wludyka and P. R. Nelson, "An Analysis of Means Type Test for Variance from Normal Populations," *Technometrics*, Vol. 39, No. 3, August 1997, pp. 274–285. © 1997 by the American Statistical Association. All rights reserved.

IN MEMORY OF PETER NELSON

Contents

Preface

The goal of statistical data analysis is to use data to gain and communicate knowledge about processes and phenomena. Comparing means is often part of an analysis, for data arising in both experimental and observational studies. Probably the most common method used to compare the means of several different treatments (or, more loosely, groups arising from stratification) is the analysis of variance (ANOVA). The analysis of means (ANOM) is an alternative procedure for comparing means. While it cannot be used in all the same settings as the ANOVA, when one is specifically interested in comparing means, such as when looking at fixed main effects in a designed experiment, ANOM has the advantages of being much more intuitive and providing an easily understood graphical result, which clearly indicates any means that are different (from the overall mean) and allows for easy assessment of practical as well as statistical significance. The graphical result is easy for nonstatisticians to understand and offers a clear advantage over ANOVA in that it sheds light on the nature of the differences among the populations.

There have been a number of advances in ANOM procedures in the last 20 years, but many of these results have appeared in fairly technical papers. ANOM is actually a multiple comparisons procedure, and the theory behind it is more complicated than that for ANOVA. Rather than dealing with univariate F distributions, one ends up with multivariate negatively correlated singular t distributions. However, the necessary critical values, power curves, and sample sizes for the ANOM procedures have already been obtained, documented and, in some instances, included in statistical software, resulting in methods that are easy to apply and with results that are easy to interpret.

Our intent in writing this book was to present the first modern comprehensive treatment of ANOM containing the information necessary for comparing means using ANOM. The book is intended to be a useful guide for practitioners, not a detailed description of the theory behind the procedures. Only as much theory as was necessary to understand and implement the various ANOM techniques is included. Most of the applications of ANOM that have appeared in the literature are from the physical sciences and engineering. However, ANOM techniques are much more broadly applicable; thus, we have included many examples from other areas, including business, medicine, health care, quality control, and the social sciences. Note that the comparison of means is used in a rather broad sense in that it also includes the comparison of Poisson rates and binomial proportions.

The audience for this book includes quality and process engineers, medical and health care investigators, social scientists, biostatisticians, epidemiologists, and scientists who may work in government, business, or education. The intended uses for this book include as a comprehensive reference for practitioners; as a text in a topics course in biostatistics, engineering statistics, industrial engineering, or business statistics; and as a supplementary text in a design of experiments course or a general course in statistical methods, health statistics, epidemiology, or biostatistics. This book is being used as a supplementary text in an introductory graduate course in statistics for health professionals, and portions of the material in this book are being used in a series of lectures given to researchers at a medical university. In addition, material in this book has been successfully used in an undergraduate course in statistical methods.

Now that ANOM is included as a standard option in some statistical software (including SAS® and MINITAB®), we anticipate the use of ANOM to expand. While not software dependent, this book includes several examples using SAS and an appendix of SAS examples that will make it easy for practitioners to implement ANOM analyses for most settings that arise in practice. While we have included many carefully worked examples that can serve as templates for practitioners who might choose to work solutions by hand, readily available software can be used to do all but the final computations; furthermore, while not explicitly illustrated, spreadsheet programs such as EXCEL™ can readily be used to perform all the calculations as well as to create ANOM decision charts.

We start with the simplest situation, in which one is interested in comparing means associated with changes in the levels of a single factor, and continue with more complicated design situations. Analysis of single-factor experiments based on the usual assumptions of normality and constant variances is covered in Chapters 2 and 3. Chapter 4 describes how to use ANOM-type techniques to test the assumption of constant variances. These chapters, together with Chapter 8, which covers analysis of normal data with nonconstant variances, and Chapter 9, which discusses distribution-free techniques, should be of interest to anyone who has means to compare. Chapter 5 discusses the ANOM for complete factorial designs, and Chapters 6 and 7 cover the ANOM for the more specialized settings of incomplete designs and axial mixture designs.

We would like to thank Robert Rodriguez for his guidance on this project and the reviewers for their insights, and most of all we wish to thank Andi Nelson for allowing us to bring this work to completion.

Peter S. Wludyka
Karen A. F. Copeland

Chapter 1

Introduction

The analysis of means (ANOM) is a graphical procedure for comparing a collection of means, rates, or proportions to see if any of them are significantly different from the overall mean, rate, or proportion. An ANOM decision chart is similar in appearance to a control chart. It has a centerline, located at the overall mean (rate or proportion), and upper and lower decision limits. The group means (rates or proportions) are plotted, and those that fall beyond the decision limits are said to be significantly different from the overall value. These differences are statistical differences, if they exist. The chart also allows one to easily evaluate the practical differences. For example, from the ANOM chart in Figure 1.1, one can conclude that the rate of office visits per member for clinic C (about 0.17 visits per member year) was significantly lower and the rate of visits for clinic A (about 0.218 visits per member year) was significantly higher than the overall rate of office visits for all clinics run by an HMO in a metropolitan area.

In circumstances in which one might use ANOVA to analyze fixed main effects, ANOM is appropriate and generally produces a more useful result. While ANOM can be used to study interactions, its main advantages occur when it is used to study main effects. When studying main effects, ANOM has two advantages over ANOVA: (1) if any of the treatments are statistically different, ANOM indicates exactly which ones are different; and (2) ANOM can be presented in a graphical form, which allows one to easily evaluate both the statistical and the practical significance of the differences.

1.1 ANOM as a Multiple Comparison Procedure

ANOM and ANOVA are only two of many ways to compare a group of means. When one is comparing exactly two means, then often the Student's t-test is used. ANOM is a graphical form of this test. For more than two means, a commonly used technique is the Tukey–Kramer (TK) procedure (Tukey (1953), Kramer (1956)) for comparing all pairwise differences of the means. There are many other multiple comparison procedures that could be used to compare means (see, e.g., Hsu (1996) or Hochberg and Tamhane (1987)). Each procedure approaches the comparison of means differently. That is, there are differences regarding exactly what is being compared. For example, ANOM compares each mean to the

1

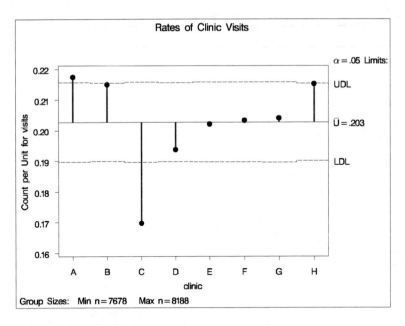

Figure 1.1. *ANOM Chart for Rates of Office Visits.*

Table 1.1. *Sample Means and Variances for Paint Drying Times (Example 1.1).*

	Paint Type			
	1	2	3	4
$\bar{y}$	6.88	9.28	9.00	9.90
s^2	1.72	1.85	2.81	1.95

overall mean, while the TK technique considers pairwise differences between the means. We examine the ANOVA, TK, and ANOM procedures in detail in the following example.

Example 1.1 (Paint Drying Data). Consider the following simple example in which one is interested in comparing the drying times (in hours) of four different types of paint used on park benches. Four benches were painted with each of the four types of paint. Summary statistics for the drying times are given in Table 1.1.

ANOVA tests for the equality of means indirectly by comparing estimates of variability that depend on the mean values. Using the ANOVA to test for differences in the drying times, one would compute the mean squares

$$\text{MS}_A = 4\{\text{sample variance of the } \bar{y}\text{s}\} = 4(1.725) = 6.9$$

and

$$\text{MS}_e = \frac{1.72 + 1.85 + 2.81 + 1.95}{4} = 2.08,$$

Table 1.2. *ANOVA Table for the Paint Drying Times (Example* 1.1).

Source	DF	Sum of Squares	Mean Square	F Value	Pr > F
Model	3	20.70250000	6.90083333	3.31	0.0572
Error	12	25.01500000	2.08458333		
Corrected Total	15	45.71750000			

where MS_A is an estimate of the variability, assuming the group means are equal, and MS_e is an estimate of variability regardless of the equality of the group means. The ratio of these two mean squares results in the test statistic

$$F = \frac{6.9}{2.08} = 3.32.$$

If there is not a significant difference in the group means, then the two measures of variability will be similar and F will be close to one. To determine statistical significance one would compare the value of F with the upper α quantile from the appropriate F distribution. Since $3.32 < F(0.05; 3, 12) = 3.49$, no significant differences are found in the drying times at level $\alpha = 0.05$. The ANOVA procedure is generally summarized in an ANOVA table, such as Table 1.2. From this table we obtain a p-value of 0.0572 for the test of equality of means. Since $\alpha = 0.05 < 0.0572$ we do not reject the hypothesis of equal means at the $\alpha = 0.05$ level.

The TK procedure considers all pairwise comparisons between group means. Using the TK procedure, one would compute simultaneous confidence intervals

$$\overline{y}_{i\bullet} - \overline{y}_{j\bullet} \pm q(\alpha; I, \nu)\sqrt{MS_e}/\sqrt{n},$$

where $q(\alpha; I, \nu)$ is the upper α quantile from a Studentized range distribution. The two means $\overline{y}_{i\bullet}$ and $\overline{y}_{j\bullet}$ are declared to be significantly different if the interval does not contain zero or, alternatively, if

$$|\overline{y}_{i\bullet} - \overline{y}_{j\bullet}| > q(\alpha; I, \nu)\sqrt{MS_e}/\sqrt{n}. \tag{1.1}$$

In our example,

$$q(0.05; 4, 12)\sqrt{MS_e}/\sqrt{n} = 4.20\sqrt{2.08}/\sqrt{4} = 3.03,$$

and the largest difference in means is $|\overline{y}_{1\bullet} - \overline{y}_{4\bullet}| = |6.88 - 9.90| = 3.02$. Thus, none of the pairs of means are different at level $\alpha = 0.05$. Table 1.3 provides computer output for the TK procedure.

Using the ANOM (details are in Chapter 2), one would compute decision lines

$$\overline{y}_{\bullet\bullet} \pm h(0.05; 4, 12)\sqrt{MS_e}\sqrt{\frac{3}{16}},$$

$$8.76 \pm 2.85\sqrt{2.08}\sqrt{\frac{3}{16}}$$

$$\pm 1.78$$

$$(6.98, 10.54).$$

Table 1.3. *TK Output for the Paint Drying Times (Example* 1.1*).*

```
Alpha                                              0.05
Error Degrees of Freedom                             12
Error Mean Square                              2.084583
Critical Value of Studentized Range             4.19852
Minimum Significant Difference                   3.0309
```

Means with the same letter are not significantly different.

Tukey Grouping	Mean	N	type
A	9.900	4	4
A	9.275	4	2
A	9.000	4	3
A	6.875	4	1

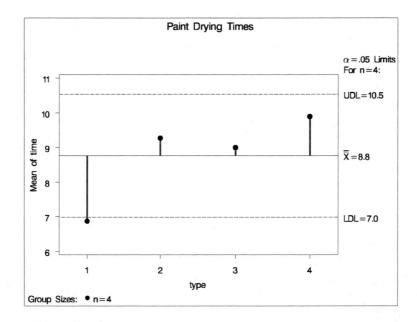

Figure 1.2. *ANOM Chart for Drying Times of Four Types of Paint (Example* 1.1*).*

From the corresponding ANOM chart in Figure 1.2, one concludes that there are significant ($\alpha = 0.05$) differences in the means because paint type 1 has a drying time that is significantly shorter than the overall average. Further, one might conclude that the difference from the overall average drying time of nearly 2 hours ($8.76 - 6.88 = 1.88$ hours) is of practical importance. Of the three procedures in this example, many practitioners will find the ANOM decision chart the easiest of the results to interpret in terms of both statistical and practical significance.

One might wonder what conclusions to draw from the fact that the three approaches to the paint example (ANOVA, TK, and ANOM) produced somewhat different results. What accounts for the different conclusions a researcher might draw? Typically, each of these are preplanned procedures and should be motivated by the purposes of the investigation. That is, what questions in particular are of interest? The three procedures just examined have somewhat different purposes and interpretations in the context of multiple comparisons. The fact that the ANOVA F test had a p-value greater than 0.05 implies that there is no Scheffé-type contrast (or set of multiple comparisons) that is significant. In particular, this implies that the set of comparisons (contrasts) in which the average for each paint is compared to the overall average is not significant at $\alpha = 0.05$ using Scheffé's multiple comparison procedure. This comparison can be made using a pair of decision lines similar to those in Figure 1.2; however, the decision lines based on Scheffé's method are wider than those used in ANOM because the Scheffé decision limits do not take into account the specific correlation structure implied by this particular set of comparisons. (Note that this structure is not a concern for ANOM users since this has been taken into account in the tables and software used in ANOM.)

ANOM uses decision lines associated with this particular set of comparisons and has associated with it exactly the level of significance specified by the test (in this case, exactly four comparisons with simultaneous significance $\alpha = 0.05$). That is, it is specifically designed to compare a group of means to the overall mean.

The TK set of pairwise comparisons is specifically designed to simultaneously test pairwise comparisons and hence is the sharpest test available for this situation. The TK ruler (the critical distance between pairs of means, which is the right-hand side of (1.1)) indicates how far apart the sample means must be to signify significance. This TK ruler typically will be less than the width (difference between the upper decision line and the lower decision line) of the ANOM decision chart. Hence, whenever at least one mean plots below the lower decision line and at least one point plots above the upper decision line, there will be at least one significant pairwise difference using TK. The central point is to use TK when pairwise comparisons will properly answer the research question.

Two Additional ANOM Examples

The following two examples illustrate the flexibility and usefulness of ANOM by showing that binomial count data and Poisson rate data can be analyzed with ANOM. These examples also illustrate how ANOM can be used in observational studies and how ANOM often provides answers to key research questions.

Example 1.2 (Epidemiological Data). A large children's clinic at a teaching hospital conducted a retrospective study of the prevalence (proportion of individuals in the population with the characteristic of interest) of obesity in the population of children they serve (predominately low-income) to determine how to package a nutritional education program. Records for the last 2 years were used to calculate the age- and sex-adjusted body mass index (BMI) percentiles for 535 children. The data were stratified by sex and race/ethnicity into six categories corresponding to sex (male or female) and race/ethnicity (black, white,

Table 1.4. *Epidemiological Study of Obesity Data (Example* 1.2*).*

Sex	Race	At Risk	Sample Size	Prevalence $= \hat{p}_i$
male	black	25	150	0.167
male	white	7	107	0.065
male	other	8	38	0.210
female	black	55	115	0.478
female	white	10	50	0.200
female	other	15	75	0.200

other) combinations. A child at or above the 85th percentile was classified as at risk for obesity. Summary data are given in Table 1.4. The research question of interest is whether the prevalence of those at risk for obesity is the same for the six strata.

One method of analysis is to perform a Pearson chi-squared test for equal prevalence, which would lead to the rejection of the hypothesis at the $\alpha = 0.01$ level. However, this sheds no light on the nature of the differences. Alternatively (or subsequently), one could examine the 15 pairwise differences, adjusting the level of significance to take into account the number of comparisons being made using, for example, the Marascuilo procedure (Marscuilo and Levin (1983)). This may not actually answer the central question since this focuses on comparing between the strata rather than among all strata, and, in addition, the large number of simultaneous comparisons reduces the power of the test.

Using the ANOM one obtains the ANOM decision chart in Figure 1.3. Prevalence for a group is judged to be different from the overall prevalence if the estimated prevalence for that group plots outside the ANOM decision lines. The decisions lines have different widths corresponding to the different sample sizes associated with each strata (wider for small samples and narrower for large samples; see Chapter 3 for details). From the decision chart one can conclude at $\alpha = 0.01$ that the prevalence of those at risk for obesity for black/female children is higher than the prevalence of those at risk for obesity in the overall clinic population. In addition, the prevalence of those at risk for obesity in the white/male population is lower than the prevalence of those at risk for obesity in the overall clinic population. Notice that due to the manner in which data were collected in this example, the overall average, $\bar{p} = (25 + 7 + 8 + 55 + 10 + 15)/535 = 0.224$, has a clear interpretation in this study as an estimate of the prevalence of at-risk children for the clinic population during the period under study. Comparing the strata (sex and race/ethnicity groups) to this has a useful interpretation and is probably more interesting than pairwise comparisons. The study strongly suggests that the nutritional educational piece be constructed to appeal to black females.

Example 1.3 (Tourism/Travel Coupon Data). A charter airline is interested in the manner in and extent to which travelers use coupons for discount opportunities at their destination. A particular traveler may use several coupons during a single travel experience. Data were collected by administering a survey to all passengers for a 2-week period. One research

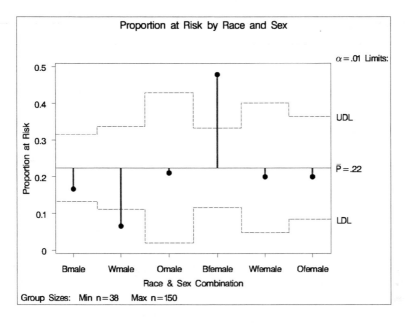

Figure 1.3. *ANOM Chart for the Obesity Data (Example 1.2).*

Table 1.5. *Coupon Use Data (Example 1.3).*

		Destination		
	Florida	Islands	New Orleans	Asheville
Passengers	525	1100	350	210
Coupons	250	505	50	260
Rate = $\hat{u}_i$	0.479	0.459	0.143	1.238

question asked whether destination affected coupon use. The survey data collected relevant to this question are summarized in Table 1.5, where the rate is the coupon use per passenger. Assuming that the rates are Poisson (see Section 3.3 for details), the ANOM decision chart for this data is shown in Figure 1.4. Since the area of opportunity (number of passengers) is different for the four destinations, the decision lines are different for each destination. In this chart, the rate for each destination is compared to the overall rate (e.g., for Florida, the rate is $\hat{u}_1 = 250/525 = 0.479$ coupons per passenger). The multiple comparison in this case consists of four comparisons, in which each of the destinations is compared to the overall rate $\bar{u} = (250 + 505 + 50 + 260)/(525 + 1100 + 350 + 210) = 0.49$, which is an estimate of the coupon use rate for all passengers. Coupon use by New Orleans passengers is significantly below average, and coupon use by those visiting Asheville is significantly higher than average.

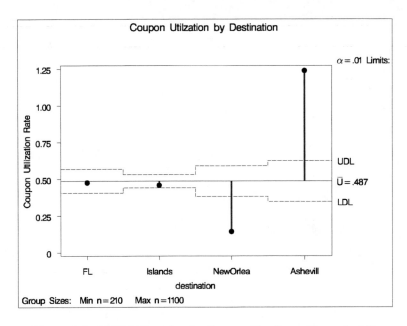

Figure 1.4. *ANOM Chart for the Coupon Use Data (Example 1.3).*

When is ANOM Most Useful?

ANOM is most useful for straightforward studies in which the desired outcome is to identify differences between groups or treatments. Similar to ANOVA, ANOM is a good choice when factor levels are clearly nominal (categorical). When factor levels are continuous, ANOM can still provide useful information; however, regression (response surface) models may be more appropriate. While ANOM has been used primarily in the analysis of experimental data, it can be very useful for analyzing observational data. This is especially true when there is post stratification, since in that circumstance the overall average estimates the population average and hence has a clear interpretation in the context of the problem (see Examples 1.2 and 1.3). Keep in mind that ANOM is a multiple comparison procedure and can be used in conjunction with other procedures. In a particular analysis, ANOM might be used in conjunction with another multiple comparison procedure, such as TK (one might wish to adjust the level of significance to control the experimentwise error rate), or in conjunction with other specific contrasts (e.g., suppose that in the coupon problem, Example 1.3, one wishes to compare tropical to nontropical destinations). A nice feature of ANOM is that the decision chart can be used to convey the conclusions arising from the data analysis. The ANOM decision chart is easy for nonstatisticians to understand. Furthermore, ANOM makes assessing practical significance easy.

1.2   History of ANOM

The basic idea of ANOM was first used by Laplace (1827), almost 100 years before Fisher (1918, 1925, 1935) introduced ANOVA. Laplace was interested in studying the homogeneity over the calendar year of the lunar atmospheric tide in Paris. He had available data on the mean change in barometric pressure from 9:00 a.m. to 3:00 p.m. over a period of 11 years. His analysis for homogeneity consisted of computing the average change for each season and comparing these with the average change over the entire year. He evaluated what we would now call the descriptive level of significance. While Laplace correctly concluded there were significant differences between the four seasons, he made what today would be considered two fundamental errors. First, he didn't account for the dependence among the four differences (seasonal averages minus the overall average), and second, rather than using the pooled seasonal variances as the measure of variability, he used the variance over the entire year.

The next appearance of an ANOM-type procedure was the multiple significance test of Halperin et al. (1955) for several normally distributed populations. They correctly used the pooled sample variances to measure variability, and to account for the dependence among the treatment means minus the overall mean, they used Bonferroni inequalities to obtain upper and lower bounds on the appropriate critical values. They conjectured that the exact critical values were closer to the lower bounds. Ott (1967) suggested applying the test of Halperin et al. (1955) in a graphical form and, based on their conjecture, provided tables of approximate critical values that were the average of the lower and upper Bonferroni bounds. Ott (1967) was the first to use the phrase "analysis of means" to describe this procedure.

Ott (1967) not only used approximate critical values but also advocated using sample ranges rather than sample variances to measure variability. Schilling (1973) extended Ott's work to designs more complicated than factorial designs with fixed effects (e.g., balanced incomplete block designs and mixed effect designs) and discussed using the ANOM with nonnormal data. Rather than using the approximate critical values proposed by Ott (1967), he used the upper bounds from Halperin et al. (1955). Schilling (1973) also advocated using sample ranges to measure variability. Following on Schilling's work, a number of authors discussed various aspects of the ANOM (Nelson (1974), Enrick (1976, 1981), Ohta (1981), Sheesley (1980, 1981)). All of this work continued to be based on conservative critical values obtained using Bonferonni's inequality.

In 1982, exact ANOM critical values for the main effects of ANOM in balanced designs were obtained (see Nelson (1982)). These exact values were based on the variability being estimated using the pooled sample variances. The entire January 1983 issue of the *Journal of Quality Technology* was devoted to ANOM and summarized the state of the art at that time. Since then there have been a number of advances in the ANOM technique. It has been shown that the exact critical values are appropriate not only for main effects from complete balanced designs but also for Latin squares, balanced incomplete block designs, Youden squares, and axial mixture designs (see Nelson (1993)). Exact critical values for main effects when the sample sizes are not equal are now available (see Nelson (1991) and Soong and Hsu (1997)), and for situations in which one doesn't want to go to the trouble of computing the exact critical values for a particular set of unequal sample sizes, conservative critical values, which are less conservative than those obtained using Bonferonni's inequality, are available (see Nelson (1989)).

Power curves for ANOM are now available (see Nelson (1985)), the ANOM technique has been extended to the case of unequal variances (see Nelson and Dudewicz (2002) and Dudewicz and Nelson (2003)), nonparametric ANOM procedures are available (see Bakir (1989) and Nelson (2002)), and the ANOM technique has even been extended to comparison of variances (see Wludyka and Nelson (1997a, 1997b, 1999) and Wludyka et al. (2001)).

Today the ANOM procedure can be found in statistical software such as SAS and MINITAB. The availability of such software has moved ANOM beyond a procedure that relies on specialized look-up tables to a technique that is easily applied in fields from engineering to managed health care.

1.3   This Book

ANOM is actually a multiple comparisons procedure (each treatment mean is compared to the overall mean), and it is only one of many possible multiple comparison procedures that could be used instead of, or as a follow up to, ANOVA. Why choose ANOM? It is easy to understand and apply, its results can be presented in a graphical form (which makes them easy to explain to those not well versed in statistics), and for practical purposes there is no disadvantage in terms of loss of power when compared with the ANOVA (see Nelson (1983a)).

One can find reference to the ANOM in a number of books (e.g., Mason et al. (1989), Farnum (1994), Vardeman (1994), Wheeler (1995), Hsu (1996), Freund and Wilson (1997), Ott et al. (2000), Ryan (2000), Nelson et al. (2003)), but there is no one place to which a practitioner can go to find simple explanations of a wide variety of the possible uses of ANOM and the necessary tables and charts. This book is intended to fill that void and presents the first modern comprehensive treatment of ANOM. While the majority of examples in the literature are of applications to problems in engineering and the physical sciences, ANOM is of much broader use. To illustrate this broad applicability, we include examples from areas such as health care, tourism, and business in addition to engineering and physical science. Further, where applicable, SAS will be used to create the ANOM charts shown in this text, and an appendix will cover the syntax needed to run the ANOM procedure in SAS by providing SAS code for some of the examples used in the book.

Chapter 2

One-Factor Balanced Studies

A simple setting often encountered is one in which an analyst is interested in comparing several (two or more) *treatments* and the same number of observations are available for each treatment. When equal numbers of observations are taken with each treatment, the experiment or study is said to be *balanced*. The treatments in an experimental design of this kind are also referred to as levels of a single *factor*. The responses recorded in a single-factor study can be either continuous or discrete data.

2.1 Types of Data

As was shown in the introduction, the ANOM procedure can be applied to continuous, proportional, and rate data. An example of each of these types of measurements follows.

Example 2.1 (Tube Weight Data). A company that produces a medical gel product packaged in 4-oz. (113.4-g) tubes recently purchased a new filling machine. One regulatory requirement is that they must validate the filling process. That is, they must demonstrate that the machine fills the tubes to the proper amount when using the machine settings that are specified for that product. As a first step in the validation process, four machine settings were studied to understand their impact on the tube fill weights. The data from seven tubes at each of the four settings are given in Table 2.1. In this example there are four treatments (different machine settings), and the first question of interest would be, Are there any differences in the machine settings with regard to the resulting fill weights? In this particular example the data are continuous and are more than likely close enough to being normally distributed that they could be modeled using a normal distribution.

Example 2.2 (On-Time Office Visit Data). An appointment is defined by a managed health company as being on time as long as the patient is taken to an exam room within 5 minutes of their scheduled appointment time. To evaluate on-time performance, 60 random appointments were selected from 1 week of records at each of 7 clinics in a geographic region. The

Table 2.1. *Fill Weights (Example* 2.1*).*

Treatment			
1	2	3	4
119.5	122.1	120.4	120.7
119.9	123.9	120.8	122.7
120.4	123.6	121.4	119.7
121.2	122.4	122.8	120.7
118.7	122.7	122.5	121.4
119.3	123.1	121.9	121.6
119.6	122.8	123.1	121.3

Table 2.2. *Number of On-Time Office Visits (Example* 2.2*).*

Clinic	A	B	C	D	E	F
On Time	10	48	37	8	35	50
Proportion On Time	0.17	0.80	0.62	0.13	0.58	0.83

on-time appointments were counted for each clinic. If a patient arrived late, the appointment was considered on time if the patient was seen within 20 minutes of arrival.

Any differences among these clinics would indicate either a clinic in need of improvement or a clinic to model improvement after. The number of on-time appointments for each clinic is given in Table 2.2. In this study, it is reasonable to model the data with a binomial distribution, since there are only two possibilities: either an appointment is on time or it is not.

Example 2.3 (Injection Molding Data). Contact lenses are manufactured in an injection molding process in which trays containing eight lens molds are injected with monomer that flows from a reservoir via eight tubes. The monomer is then cured by exposure to UV light. The cure time is determined by the speed at which a conveyor belt moves. The process engineer wants to know the extent to which belt speed (cure time) influences the incidence of small nonconformities (such as bubbles and tears) in the lenses. A particular lens may have $0, 1, 2, \ldots$ nonconformities. To investigate the relationship between belt speed and the incidence of nonconformities, 100 trays of lenses were produced at each of five belts speeds (measured in feet/second), and the total number of nonconformities was recorded for each speed. The data are given in Table 2.3. The data in this example can be modeled with a Poisson distribution since in a very small lens area there would be at most one nonconformity, and the total number of nonconformities could be arbitrarily large.

Table 2.3. *Nonconformities for Five Belt Speeds (Example* 2.3*).*

Speed	10	15	20	25	30
Nonconformities	63	48	69	82	124

We will consider how to apply ANOM to test for differences with each of these types of data.

2.2 Normally Distributed Data

With data that are at least approximately normally distributed, one can test for differences in the treatments using the sample means and sample variances for the different treatments. Since there will be several sample means and sample variances, we need to introduce some notation.

Notation. We will use y_{ij} to represent the jth observation of the ith treatment. Let I be the total number of treatments (factor levels), and let n be the number of observations with each treatment. We will use dot notation to indicate averaging. For example,

$$\overline{y}_{i\bullet} = \frac{1}{n} \sum_{j=1}^{n} y_{ij}$$

= the average of the responses for the ith treatment.

In addition,

$$s_i^2 = \frac{1}{n-1} \sum_{j=1}^{n} (y_{ij} - \overline{y}_{i\bullet})$$

= the variance of the responses for the ith treatment,

$$\overline{y}_{\bullet\bullet} = \frac{1}{N} \sum_{i=1}^{I} \sum_{j=1}^{n} y_{ij} = \frac{1}{I} \sum_{i=1}^{I} \overline{y}_{i\bullet} = \text{the overall mean,}$$

N = the total number of observations.

Example 2.4 (Fill Weight Data). To illustrate this notation, all observation and the sample means and variances for Example 2.1 have been carefully labeled in Table 2.4. In this example, we have $I = 4$ treatments with $n = 7$ *replicates* (i.e., observations per treatment), $N = 28$ total observations, and an overall mean of

$$\overline{y}_{\bullet\bullet} = \frac{\overline{y}_{1\bullet} + \cdots + \overline{y}_{4\bullet}}{4}$$

$$= \frac{119.8 + 122.9 + 121.8 + 121.2}{4} = 121.4.$$

Table 2.4. *Illustration of the Notation for the Data (Example 2.1).*

	Treatment		
1	2	3	4
$y_{11} = 119.5$	$y_{21} = 122.1$	$y_{31} = 120.4$	$y_{41} = 120.7$
$y_{12} = 119.9$	$y_{22} = 123.9$	$y_{32} = 120.8$	$y_{42} = 122.7$
$y_{13} = 120.4$	$y_{23} = 123.6$	$y_{33} = 121.4$	$y_{43} = 119.7$
$y_{14} = 121.2$	$y_{24} = 122.4$	$y_{34} = 122.8$	$y_{44} = 120.7$
$y_{15} = 118.7$	$y_{25} = 122.7$	$y_{35} = 122.5$	$y_{45} = 121.4$
$y_{16} = 119.3$	$y_{26} = 1123.1$	$y_{36} = 121.9$	$y_{46} = 121.6$
$y_{17} = 119.6$	$y_{27} = 122.8$	$y_{37} = 123.1$	$y_{47} = 121.3$
$\bar{y}_{1\bullet} = 119.8$	$\bar{y}_{2\bullet} = 122.9$	$\bar{y}_{3\bullet} = 121.8$	$\bar{y}_{4\bullet} = 121.2$
$s_1^2 = 0.653$	$s_2^2 = 0.410$	$s_3^2 = 1.05$	$s_4^2 = 0.866$

Assumptions

With continuous data, the ANOM procedure is based on two assumptions:

(i) The data are at least approximately normally distributed.

(ii) The different treatments all have the same variance.

> **Notation.** We will use the standard notation of representing random variables with uppercase letters and the specific values of these random variables with lowercase letters. Thus, when we are talking about the distribution of the observations, we will refer to Y_{ij}, and y_{ij} is an observed value of Y_{ij}.

We could write these assumptions more compactly as

$$Y_{ij} \sim N(\mu_i, \sigma^2),$$

which is read "Y_{ij} has a normal distribution with mean μ_i and variance σ^2." This could be rewritten as

$$\boxed{Y_{ij} = \mu_i + \epsilon_{ij}}, \tag{2.1}$$

where

$$\mu_i = \text{the mean for the } i\text{th treatment,}$$
$$\epsilon_{ij} = \text{the random error associated with } Y_{ij},$$

and the assumptions of normality and equal variances imply $\epsilon_{ij} \sim N(0, \sigma^2)$. Using model (2.1), one can check the normality assumption with a single normal probability plot since the ϵ_{ij} all should have the same normal distribution. If one looked directly at the y_{ij}, one would need a different plot for each treatment because the means (i.e., μ_i) are potentially different.

The ϵ_{ij} are random variables, but we can compute the specific errors associated with the observed y_{ij}. The μ_i can be estimated by the sample means

$$\widehat{\mu}_i = \overline{y}_{i\bullet},$$

and (2.1) can be rearranged to give

$$\widehat{\epsilon}_{ij} = y_{ij} - \overline{y}_{i\bullet} . \tag{2.2}$$

Notation. The symbol $\widehat{\mu}_i$ (read "mu sub i hat") means an estimate of μ_i.

The $\widehat{\epsilon}_{ij}$ are referred to as *residuals*. Note that the residuals are really nothing more than the original observations shifted so that each set of treatment values is centered at zero. Both the assumptions of normality and the equal variances can be checked graphically, to rule out any obvious problems, using a normal probability plot of the residuals. A normal probability plot is actually a quantile/quantile plot in which the sample quantiles from the data are plotted against the quantiles for a normal distribution. If the plot looks approximately linear, then the normal distribution is a reasonable model for the data. There are several different rules about what should be used for the sample quantiles, but one of the most common is that $x_{(i)}$ (the ith smallest data value) is the $p_i = \frac{i-0.5}{N}$ quantile, where N is the total number of data values. One plots the ordered pairs $(Q(p_i), x_{(i)})$, where $Q(p_i)$ is the p_ith quantile from a standard normal distribution. This is most easily done by using statistical software with a normal probability plotting capability.

Example 2.5 (Customer Lifetime Data). A credit card transaction processing company serves customers from various retail sectors. To study how the lifetimes of customers differ between sectors, the lifetimes, or the length of time a customer remains a customer, were calculated for 10 customers for each of three sector types (convenience stores, beauty salons, and medical offices). The data are recorded in Table 2.5. A normal probability plot to check the assumptions is given in Figure 2.1, from which it is clear that the residuals are not approximately normally distributed, and it would therefore not be appropriate to apply the ANOM to these data. However, if one transforms the data using a log transformation (i.e., $x \to \ln(x)$), the data then look at least approximately normal, and it would be appropriate to analyze the transformed data using ANOM (see Problem 2.1).

Example 2.6 (Random Normal Data). The data in Table 2.6 are random samples from three normal populations. Population 1 is $N(25, 0.01)$, population 2 is $N(5, 25)$, and population 3 is $N(0, 100)$. The vast differences in the sample variances would suggest that the variances are not all equal. This is confirmed by the normal probability plot given in Figure 2.2, where the data points fall in three lines with different slopes, which correspond to the three different variances.

Table 2.5. *Lifetimes of Customers (Example 2.5).*

Sector		
Convenience Stores	Beauty Salons	Medical Offices
0.5	1.3	8.5
0.7	1.9	12.5
0.9	5.3	5.2
1.3	12.5	7.2
5.0	0.6	8.5
0.6	0.5	15.8
1.5	6.4	21.4
1.7	0.1	25.3
7.3	2.1	5.6
10.2	0.8	6.2

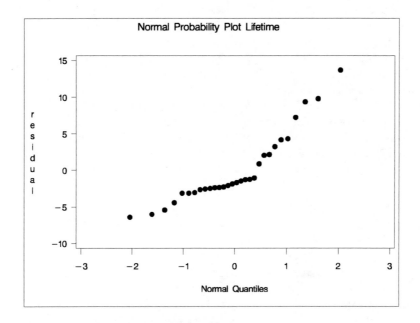

Figure 2.1. *Normal Probability Plot of the Residuals (Example 2.5).*

Example 2.7 (Tube Weight Data). A normal probability plot for this set of data is given in Figure 2.3, from which one sees that the assumptions of normality and equal variances appear to be reasonable since the data fall in a straight line.

ANOM

Once we are satisfied that the assumptions of normality and equal variances are reasonable, then we can test to see whether the factor has any effect. Since we have assumed the different

Table 2.6. *Random Samples from Three Normal Populations (Example 2.6).*

	Population	
1	2	3
25.07	4.91	2.19
24.96	−3.27	2.40
24.99	5.43	17.70
25.01	5.68	0.58
25.03	3.89	2.80
24.77	3.50	−13.62
24.95	7.88	6.21
24.99	3.47	−4.88
$s_1^2 = 0.008$	$s_2^2 = 10.58$	$s_3^2 = 79.65$

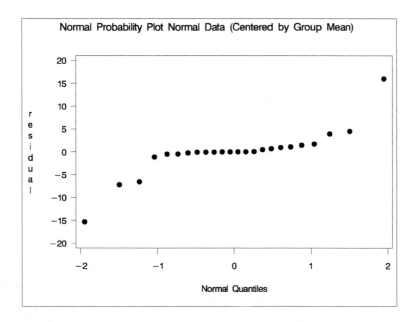

Figure 2.2. *Normal Probability Plot for the Data (Example 2.6).*

levels of the factor have the same variance, we need only check if the factor has an effect on the means. That is, one would test

$$H_0 : \mu_1 = \mu_2 = \cdots = \mu_I$$

versus the alternative hypothesis that at least one of the μ_i is different.

Using the ANOM to test this hypothesis not only answers the question of whether there are any differences among the factor levels but, as we will see below, when there are differences, it also indicates how the treatment levels differ. Determining differences with

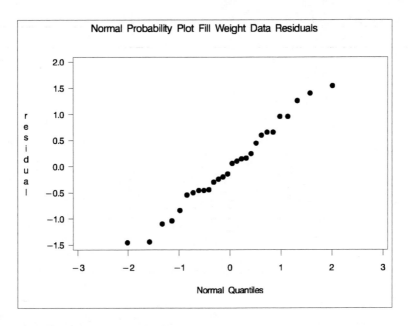

Figure 2.3. *Normal Probability Plot for the Tube Fill Weight Data (Example 2.1).*

ANOM is facilitated by a decision chart that is easy to understand and can be used to explain results to nonstatisticians.

The idea of the analysis of means is that if H_0 is true, the I factor levels all have the same population mean. Therefore, all the $\overline{y}_{i\bullet}$ should be close to the overall mean $\overline{y}_{\bullet\bullet}$. So we will reject H_0 if any one of the $\overline{y}_{i\bullet}$ is too far away from $\overline{y}_{\bullet\bullet}$. In the discussion of ANOM that follows, we will assume that all the factor levels have equal sample sizes. The case in which the sample sizes are unequal is slightly more complicated and is covered in the next chapter.

How far away is too far depends on the common variance, which can be estimated by

$$\widehat{\sigma}^2 = \mathrm{MS}_e = \frac{s_1^2 + \cdots + s_I^2}{I} \ . \tag{2.3}$$

The quantity $\widehat{\sigma}^2$ is referred to as the *mean square error* (MS_e), and in this case, which has only a single factor, it has $N - I$ degrees of freedom (df).

The ANOM is performed by computing upper and lower decision lines,

$$
\begin{aligned}
\mathrm{UDL} &= \overline{y}_{\bullet\bullet} + h(\alpha; I, N - I)\sqrt{\mathrm{MS}_e}\sqrt{\frac{I-1}{N}} \\[2mm]
\mathrm{LDL} &= \overline{y}_{\bullet\bullet} - h(\alpha; I, N - I)\sqrt{\mathrm{MS}_e}\sqrt{\frac{I-1}{N}}
\end{aligned}
\tag{2.4}
$$

and checking to see if any of the treatment means (i.e., factor level means) fall outside these decision lines by plotting the treatment means with the decision lines. The critical values $h(\alpha; I, \nu)$ depend on

$$\alpha = \text{the level of significance desired,}$$
$$I = \text{the number of means being compared,}$$
$$\nu = \text{the degrees of freedom for MS}_e,$$

and are given in Table B.1 in Appendix B. They are derived based on the joint distribution of the $|\bar{y}_{i.} - \bar{y}_{..}|$, which is an equicorrelated multivariate t distribution with correlations $\rho = -1/(I - 1)$. (See Nelson (1982, 1993).)

Example 2.8 (Tube Fill Weight). Consider the four filling machine settings again. The four treatment averages and the overall average are (from Example 2.4)

$$\bar{y}_{1.} = 119.8, \quad \bar{y}_{2.} = 122.9, \quad \bar{y}_{3.} = 121.8, \quad \bar{y}_{4.} = 121.2, \quad \text{and} \quad \bar{y}_{..} = 121.4.$$

The sample variances (also from Example 2.4) are averaged (equation (2.3)) to obtain

$$\text{MS}_e = \frac{0.653 + 0.410 + 1.05 + 0.866}{4} = 0.745$$

with $N - I = 28 - 4 = 24\,$df. For $\alpha = 0.01$ one finds, from Table B.1, $h(0.01; 4, 24) = 3.35$ and

$$h(0.01; 4, 24)\sqrt{\text{MS}_e}\sqrt{\frac{I-1}{N}} = 3.35\sqrt{0.745}\sqrt{\frac{3}{28}} = 0.94.$$

Thus,
$$\text{UDL} = 121.4 + 0.94 = 122.34,$$
$$\text{LDL} = 121.4 - 0.94 = 120.46,$$

and one obtains the ANOM chart in Figure 2.4.

From the chart it is clear that there is a significant effect (at $\alpha = 0.01$) due to machine setting 1 under filling tubes and machine setting 2 over filling tubes when compared to the average tube weight.

The ANOM chart also provides an easy way to assess *practical significance*. That is, once we have found that the treatments are statistically different, one can ask if there are really any practical differences. In the case of the machine settings, on average setting 1 results in $121.4 - 119.8 = 1.6$ g less of material in the tube, and setting 2 results in $122.9 - 121.4 = 1.5$ g more of material. One fill run takes about 2 hours and fills roughly 3000 tubes. Thus, the differences is about 40 tubes of product (more for setting 1 or less for setting 2) per fill run. Note that practical significance is not a statistical issue, and it makes sense to ask about practical significance only if one first finds statistical significance, since without statistical significance one can't distinguish between the treatments.

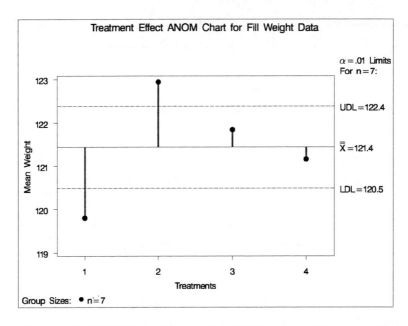

Figure 2.4. *ANOM Chart for the Tube Fill Weight Data (Example 2.8).*

Example 2.9 (Injection Depth Data). In the development of a needleless injector, a medical device company needed to study the performance of its device when used at various injection sites. Three sites were chosen; the buttocks, the thigh, and the post thigh. Four injections were made at each site, and the depth of each injection was measured. (Cadavers were used for this study.) Of interest was whether there were differences in the injection depth at various sites. To begin, a normal probability plot of the residuals (Figure 2.5) is used to show that the assumptions of normality and equal variances are reasonable since the data on the plot fall in a straight line (indicating that they come from a normal distribution with the same variances). Table 2.7 contains the data and summary statistics. From these values one calculates

$$\mathrm{MS}_e = \frac{1.242 + 1.537 + 3.342}{3} = 2.04$$

with $N - I = 12 - 3 = 9$ df. For $\alpha = 0.05$ one finds $h(0.05; 3, 9) = 2.79$ and

$$h(0.05; 3, 9)\sqrt{\mathrm{MS}_e}\sqrt{\frac{I-1}{N}} = 2.79\sqrt{2.04}\sqrt{\frac{2}{12}} = 1.63.$$

Thus, the decision lines are given by

$$\mathrm{UDL} = 18.28 + 1.63 = 19.91,$$
$$\mathrm{LDL} = 18.28 - 1.63 = 16.65.$$

The resulting ANOM chart in Figure 2.6 clearly shows that the injection depth is significantly greater for injections in the thigh than the average injection depth. If a deeper injection is

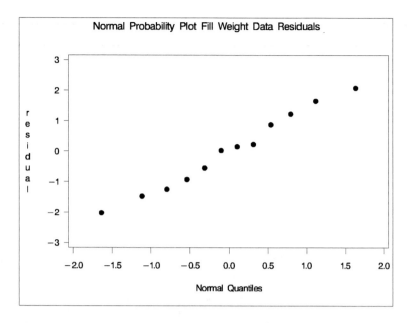

Figure 2.5. *Normal Probability Plot of Residuals for Injection Depth Data (Example 2.9).*

Table 2.7. *Data and Summary Statistics for Injection Depth Data (Example 2.9).*

	Injection Site	
Buttocks	Post Thigh	Thigh
17.3	18.7	22.6
18.5	15.8	19.6
17.5	16.5	21.4
15.8	17.2	18.5
$\bar{y}_{1\bullet} = 17.275$	$\bar{y}_{2\bullet} = 17.050$	$\bar{y}_{3\bullet} = 20.525$
$s_1^2 = 1.242$	$s_2^2 = 1.537$	$s_3^2 = 3.342$

better, then the company can include in the product packet insert the recommendation that the injection be given in the thigh.

Example 2.10 (Statistical Process Control Data). Statistical process control (SPC) often involves two stages. In stage I, data are collected from the process with the initial goal of determining whether the process is in control. If so, then in-control data points are used to set up a control scheme. If not, steps are taken to get the process under control. Stage I is sometimes characterized as establishing retrospective control. In-control data points are then used prospectively to operate a control scheme such as a k-sigma Shewhart chart. In this example a quality engineer wishes to control the diameter of cups used in an injection

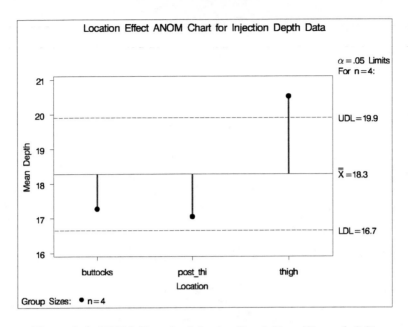

Figure 2.6. *ANOM Chart for Injection Depth Data (Example 2.9).*

molding process with a 3-sigma $\overline{X}$ control chart. To this end she collects data every 30 minutes from the production line. The data, given in Table 2.8, consist of measurements of 5 cups from each of 10 samples. The measurements are the deviation from target diameter in millimeters. Taking the deviation from the target simply shifts the data by the target value. One could as easily perform the following analysis on the actual values. One approach to establishing retrospective control is to plot the 10 sample means on a 3-sigma $\overline{X}$ chart, using that to gauge whether the process is under control. Recall that in this chart the sample means are plotted on a chart with control limits and centerline determined by the data associated with the points being plotted. An alternative for establishing retrospective control is to use an ANOM decision chart to test whether the means are equal (and hence observations from an in-control process). This is suitable since ANOM explicitly takes into account that the grand mean and the sample means are correlated. (This dependence is ignored in the standard control chart.) The ANOM decision lines as shown in Figure 2.7 with $\alpha = 0.01$ are

$$-0.052 \pm h(0.01;\ 10,\ 40)\sqrt{0.829}\sqrt{\frac{9}{50}}$$
$$\pm\ 3.54(0.491)$$
$$\pm\ 1.366$$
$$(-1.419,\ 1.315)$$

and indicate that the process is not under control.

Suppose that the quality engineer has taken steps to bring the process under control. Subsequently, 20 samples were collected from the production line, resulting in the decision

Table 2.8. *Deviations from Target Diameter (SPC) Data (Example* 2.10).

Sample	Deviations				
	1	2	3	4	5
1	1.745	0.102	−0.534	−0.934	0.232
2	1.162	1.608	2.089	2.063	3.283
3	3.454	1.624	1.590	−0.330	2.440
4	0.355	0.332	−1.044	−0.397	0.181
5	−1.002	−1.042	−0.900	−0.295	0.153
6	0.856	−0.388	0.787	−1.223	−1.012
7	−1.103	0.505	0.421	0.610	−0.708
8	−2.938	−1.821	−2.277	−2.306	−1.554
9	0.939	2.330	−0.339	0.028	1.163
10	−3.381	−2.439	−1.862	−2.300	−0.532

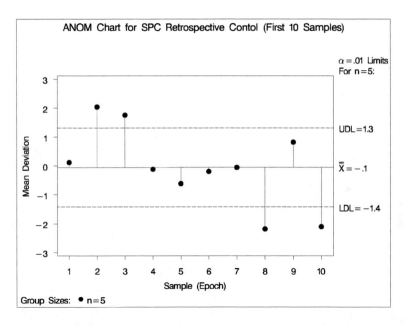

Figure 2.7. *ANOM Chart for Deviations from Target (Example* 2.10).

chart in Figure 2.8. The sample mean for epoch 3 plotted above the upper ($\alpha = 0.05$) decision line, indicating that during that epoch the process was out of control. A search for an assignable cause determined that the filling gauge had malfunctioned. That point was dropped from the data set and a new ANOM decision chart was constructed (Figure 2.9). This time, all the points plotted within the decision limits, indicating that the process had been in control during those 19 epochs. Estimates of the process mean and standard deviation derived from this data could then be used to create a stage II $\overline{X}$ chart for maintaining process control. It should be noted that for 20 or so samples, the decision limits for a retrospective

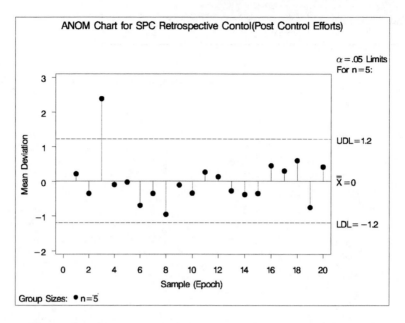

Figure 2.8. *ANOM Chart for Deviations from Target (Example 2.10).*

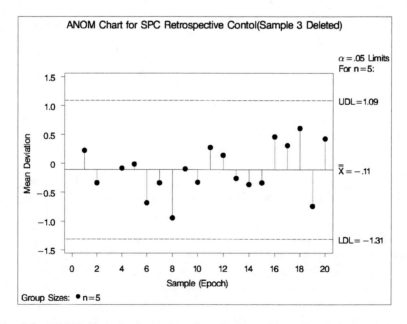

Figure 2.9. *ANOM Chart for Deviations from Target without Epoch 3 (Example 2.10).*

3-sigma $\overline{X}$ chart will be very similar to those for the $\alpha = 0.05$ ANOM decision chart. The advantage offered by the latter is that the Type II error rate is known to be 5% (or whatever level of significance is specified).

ANOM p-Values

Researchers often wish to cite the p-value, or the descriptive level of significance, associated with a test (denoted as p in this section). Citing p-values is particularly common in abstracts. (The practice is facilitated by software packages, where, for example, using the data from the injection depth penetration study (Example 2.9), the ANOVA F test has $p = 0.0125$.) The p-value is the probability, when H_0 is true, of obtaining a test statistic as extreme, or more extreme, than the one actually obtained. Small p-values lead to the rejection of H_0. Note that if a hypothesis is rejected at level of significance α, then $p < \alpha$. So, from the ANOM chart for $\alpha = 0.05$ in the injection study (see Figure 2.6), one can conclude that for the ANOM procedure $p < 0.05$ since at least one point plots outside the $\alpha = 0.05$ decision limits. Using table look-up methods (and calculating the corresponding decision limits), one can ascertain that in this example $0.005 < p < 0.01$ since one point plots outside the $\alpha = 0.01$ decision limits and all points plot within the $\alpha = 0.005$ decision limits. For most purposes a statement of this type will suffice. A more precise value for p can be determined using the notion of implied level of significance as a definition of p. Then, in the balanced case, one can identify the p-value for ANOM as the value of α such that the most extreme point (the one with the greatest distance from the centerline) coincides with one of the decision lines. So, for the injection depth data, $p = 0.0099$, since in the decision chart in Figure 2.10 the mean depth for a thigh injection coincides with the upper decision line (UDL $= 20.525$) for $\alpha = 0.0099$, and all other points plot within or on the decision lines. (Observe that the p-value for the ANOVA F test is not very different from the p-value for the ANOM procedure.) More precisely, p is determined by solving the equation

$$h(p; I, N - I) = \sqrt{\frac{N}{I-1}} \sqrt{\frac{1}{MS_e}} \max |\overline{Y}_{i\bullet} - \overline{Y}_{i\bullet\bullet}|.$$

In practice, when more than an interval for p is desired, one can use a package such as SAS and iteratively (trial and error is sufficient) determine p by running the ANOM procedure until the most extreme point (mean) coincides with one of the decision lines.

Sample Sizes

It is possible to obtain sample sizes for the ANOM if one specifies both the level of significance and the power. One has to specify the power for a particular configuration of unequal means. Generally, one measures how far a configuration of unequal means is from the equal means configuration by the maximum distance between any two means. That is, one specifies the power in terms of

$$\Delta = \max_{i,j} \frac{|\mu_i - \mu_j|}{\sigma}.$$

Since σ is rarely known, one simply specifies a value for Δ, which can be thought of as the maximum difference between any two means (measured in units of σ) before the means are

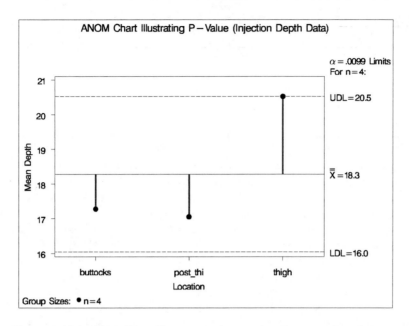

Figure 2.10. *ANOM Chart Illustrating the p-value (Injection Depth Data).*

far enough apart that you want to reject H_0. For a given α-level, power, specified detectable difference, and number of treatment groups, the number of observations that is needed in each treatment group is given in Table B.2. (See Nelson (1983a) for a discussion of how these tables were generated.) For more precise power considerations, one can use power curves (see Nelson (1985) for details). Power curves for $\alpha = 0.05$ and $k = 3 - 8$, 10, and 12 are given in Appendix A.

Example 2.11 (Injection Depth Data). Suppose that in the injection depth penetration study (Example 2.9), the experimenters needed to detect a 3σ difference between any two injection sites with probability 0.90 when testing using the ANOM and $\alpha = 0.05$. How many injections would have to be made at each site?

From Table B.2 with $\alpha = 0.05$, $\Delta = 3$, power $= 0.9$, and $k = 3$, one obtains $n = 5$. Thus five injections would be needed at each site. However, the study was run with only four injections at each site. What is the power of the current study to detect a 3σ difference in means using the ANOM with $\alpha = 0.05$? Using Table B.2 in reverse, one finds that with $\alpha = 0.05$, $\Delta = 3$, $k = 3$, and $n = 4$, the power is 0.80. From Figure A.3 one finds that for $n = 4$ the power is about 0.88.

Example 2.12 (Fill Weight Data). The company in Example 2.1 has used its new machine for a number of fill runs and would now like to design a new study of the four machine settings such that they can detect a difference in fill weights of 1.5g with power $= 0.80$ using the ANOM with $\alpha = 0.01$. How many tubes have to be filled on each setting? The company is willing to use 0.75 (based on the MS_e from the previous study) as an estimate of σ.

From Table B.2 with $\alpha = 0.01$, $\Delta = (1.5)/(0.75) = 2$, power $= 0.80$, and $k = 3$, one obtains $n = 9$, so nine tubes must be filled at each setting.

2.3 Binomial Data (ANOM for Proportions)

The ANOM procedure described in the last section is based on the assumption that the observations are normally distributed. That procedure is easily modified to allow for comparisons of several proportions when the sample sizes are large enough to use the normal approximation for the binomial distribution. To compare several proportions, we will test the hypothesis

$$H_0 : p_1 = \cdots = p_I$$

against the alternative that at least one of the p_i is different. When the proportions are all based on equal sample sizes, then one modifies the decision lines (2.4). One can think of the sample proportion $\widehat{p}_i$ from a sample of size n as being the average of n Bernoulli random variables. That is,

$$\widehat{p}_i = \frac{1}{n} \sum_{j=1}^{n} x_{ij},$$

where

$X_{ij} = 1$ if the jth item from the ith group is nonconforming,

$\quad\quad = 0$ otherwise.

> **Note.** We use the "nonconforming" terminology when referring to Bernoulli random variables, but in general the Bernoulli model applies to any situation in which the items sampled can be put into one of two categories. Two other possibilities are male/female and pass/fail.

The x_{ij} are the counterparts to the y_{ij} used in the decision line (2.4). The variance of X_{ij} is $p_i(1 - p_i)$, which under the null hypothesis does not depend on i. Thus, $p(1 - p)$ is the counterpart to σ^2, and the best estimator of that quantity (assuming the null hypothesis is true) is $\overline{p}(1 - \overline{p})$, where

$$\overline{p} = \frac{\widehat{p}_1 + \cdots + \widehat{p}_I}{I} . \tag{2.5}$$

The degrees of freedom associated with this estimator are ∞ since we are using the normal approximation. Taking all this into account results in the decision lines

$$
\begin{aligned}
\text{UDL} &= \overline{p} + h(\alpha; I, \infty)\sqrt{\overline{p}(1 - \overline{p})}\sqrt{\tfrac{I-1}{N}} \\[2mm]
\text{LDL} &= \overline{p} - h(\alpha; I, \infty)\sqrt{\overline{p}(1 - \overline{p})}\sqrt{\tfrac{I-1}{N}}
\end{aligned}
\tag{2.6}
$$

To be sure the sample size is large enough to use the normal approximation to the binomial (and, therefore, the decision lines (2.6)), one would check that $n\widehat{p}_i > 5$ and $n(1 - \widehat{p}_i) > 5$ for all i. It suffices to check these conditions for just

$$\widehat{p}_{\min} = \text{the smallest } \widehat{p}_i$$

and
$$\widehat{p}_{\max} = \text{the largest } \widehat{p}_i.$$

In fact, one needs to check only that

$$n\widehat{p}_{\min} > 5 \quad \text{and} \quad n(1 - \widehat{p}_{\max}) > 5 . \tag{2.7}$$

Example 2.13 (On-Time Office Visit Data). In Example 2.2, the sample proportions are

$$\widehat{p}_1 = 0.167, \quad \widehat{p}_2 = 0.800, \quad \widehat{p}_3 = 0.617, \quad \widehat{p}_4 = 0.133, \quad \widehat{p}_5 = 0.583, \quad \widehat{p}_6 = 0.833,$$

and

$$\overline{p} = \frac{0.167 + \cdots + 0.833}{6} = 0.522.$$

Checking the conditions (2.7) for this example, we have

$$n\widehat{p}_{\min} = 60(0.133) = 7.98 > 5,$$

$$n(1 - \widehat{p}_{\max}) = 60(1 - 0.833) = 60(0.167) = 10.02 > 5.$$

Thus, the sample size is large enough, and the ANOM decision lines (2.6) with $\alpha = 0.01$ are

$$0.522 \pm h(0.01; 6, \infty)\sqrt{0.522(1 - 0.522)}\sqrt{\frac{5}{360}}$$

$$\pm 3.14(0.0589)$$

$$\pm 0.185$$

$$(0.337, 0.707).$$

The ANOM chart is given in Figure 2.11, from which one sees that clinics A and D had significantly low on-time performance, while clinics B and F had significantly high on-time performance, as compared to the overall clinic mean of 0.522.

Example 2.14 (Tube Seal Data). The filling machine discussed in Example 2.1 also seals the tubes after they have been filled. There are three heat settings for the sealer (low, medium, and high). Fifty tubes were sealed on each setting, and the number of tubes that burst under extreme pressure is recorded in Table 2.9. Checking conditions (2.7), we find that the sample size is large enough to use the ANOM since

$$n\widehat{p}_{\min} = 50(0.12) = 6 > 5,$$

$$n(1 - \widehat{p}_{\max}) = 50(1 - 0.24) = 50(0.76) = 38 > 5.$$

The ANOM decision lines (2.6) with $\alpha = 0.05$ are

$$0.18 \pm h(0.05; 3, \infty)\sqrt{0.18(1 - 0.18)}\sqrt{\frac{2}{150}}$$

$$\pm 2.34(0.044)$$

$$\pm 0.104$$

$$(0.076, 0.284).$$

The ANOM plot in Figure 2.12 shows that at the 0.05 significance level there is not a significant effect of temperature on tube seal strength.

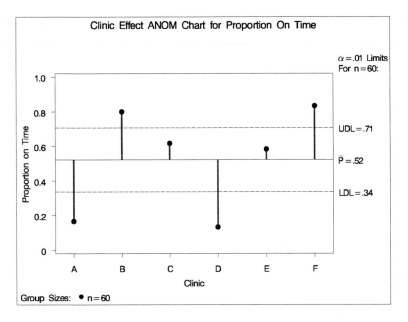

Figure 2.11. *ANOM Chart for the On-Time Office Visit Data (Example* 2.13*).*

Table 2.9. *Number of Burst Tubes (Example* 2.14*).*

Temperature	Low	Medium	High
Number	12	6	9
Proportion	0.24	0.12	0.18

2.4   Poisson Data (ANOM for Frequencies or Rates)

The ANOM procedure for normally distributed data can be modified to allow for comparisons of several frequencies when the sample sizes are large enough to use the normal approximation to the Poisson distribution. To compare several frequencies, we will test the hypothesis

$$H_0 : u_1 = \cdots = u_I$$

against the alternative that at least one of the u_i is different. When the frequencies are all based on equal sample sizes, one modifies the decision lines (2.4). When the null hypothesis is true, both the mean and the variance of the single Poisson distribution would be estimated with the average frequency

$$\bar{u} = \frac{\widehat{u}_1 + \cdots + \widehat{u}_I}{I}. \tag{2.8}$$

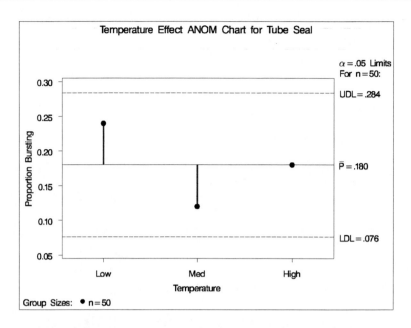

Figure 2.12. *ANOM Chart for the Tube Seal Data (Example* 2.14).

Since we are again using the normal approximation, the df for the estimated variance are infinite and the decision lines are

$$
\begin{aligned}
\text{UDL} &= \bar{u} + h(\alpha; I, \infty)\sqrt{\bar{u}}\sqrt{\tfrac{I-1}{N}} \\
\text{LDL} &= \bar{u} - h(\alpha; I, \infty)\sqrt{\bar{u}}\sqrt{\tfrac{I-1}{N}}
\end{aligned}
\tag{2.9}
$$

To ensure that the sample sizes are large enough to use the normal approximation to the Poisson, one would check that $n\widehat{u}_i > 5$ for all i. To do this it would suffice to check

$$
n\widehat{u}_{\min} > 5 .
\tag{2.10}
$$

Example 2.15 (Urgent Care Arrival Data). An HMO runs six urgent care clinics that each serve about the same number of members. To compare the arrival rates of patients at these six clinics it was decided to record the number of patients served on each of five preselected days (the same five days were used at all clinics). Each clinic is open for 12 hours per day, so the resulting data were the number of patients served during 60 hours of service time (the opportunity time). The data are given in Table 2.10. The average frequency is (equation (2.8))

$$
\bar{u} = \frac{6.217 + \cdots + 8.3}{6} = 6.945.
$$

Table 2.10. *Arrival Rates at Urgent Care Clinics (Example 2.15).*

Clinic	A	B	C	D	E	F
Arrivals	373	347	465	388	429	498
Rate (per hour)	6.217	5.783	7.75	6.467	7.15	8.3

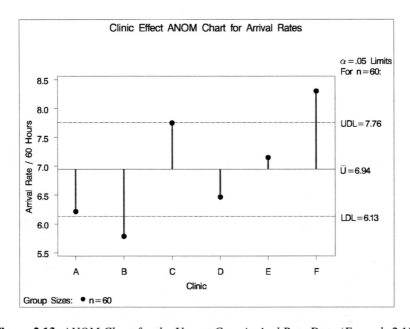

Figure 2.13. *ANOM Chart for the Urgent Care Arrival Rate Data (Example 2.15).*

Since $60(5.783) = 347 > 5$, the sample sizes are large enough to use the normal approximation to the Poisson. Thus, the $\alpha = 0.05$ ANOM decision lines (2.9) are

$$6.945 \pm h(0.05; 6, \infty)\sqrt{6.945}\sqrt{\frac{5}{360}}$$

$$\pm 2.62(0.311)$$

$$\pm 0.815$$

$$(6.13, 7.76).$$

From the ANOM chart given in Figure 2.13, one sees that there are significant ($\alpha = 0.05$) differences because clinic B has a low arrival rate of patients and clinics C and F have high arrival rates.

Example 2.16 (Injection Molding Data). In Example 2.3 the sample frequencies are

$$\widehat{u}_1 = 0.63, \quad \widehat{u}_2 = 0.48, \quad \widehat{u}_3 = 0.69, \quad \widehat{u}_4 = 0.82, \quad \widehat{u}_5 = 1.24,$$

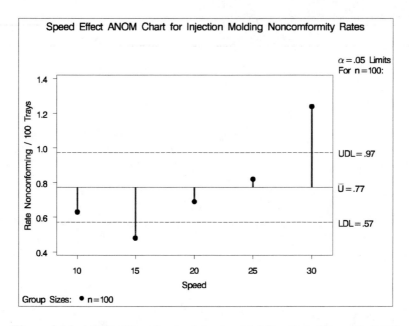

Figure 2.14. *ANOM Chart for the Injection Molding Data (Example 2.16).*

and the average frequency is (equation (2.8))

$$\bar{u} = \frac{0.63 + \cdots + 1.24}{5} = 0.772.$$

Since $100(0.48) = 48 > 5$, the sample sizes are large enough, and the $\alpha = 0.05$ ANOM decision lines (2.9) are

$$0.772 \pm h(0.05; 5, \infty)\sqrt{0.772}\sqrt{\frac{4}{500}}$$

$$\pm 2.56(0.0786)$$

$$\pm 0.201$$

$$(0.57, 0.97).$$

From the ANOM chart given in Figure 2.14, one sees that there are significant ($\alpha = 0.05$) differences due to a belt speed of 15 resulting in a significantly low number of nonconformities and a belt speed of 30 resulting in a significantly high number of nonconformities.

Chapter 2 Problems

1. Transform the customer lifetime data in Example 2.5 using a log transformation, make a normal probability plot of the transformed data to show that the transformation was successful, and analyze the transformed data using ANOM.

2. A finish is applied to the tops of small decorative tables. Dust and other particulate matter in the air in the drying room are thought to contribute to the formation of flaws (such as bubbles) on the surface. To reduce particulate matter, the dry room is evacuated by fans before to each drying. To determine how long to evacuate the room, an experiment was conducted using evacuation times of 5 minutes, 10 minutes, and 15 minutes. Fifty tables were dried for each evacuation time, and the results are recorded below.

	Evacuation Time		
	5	10	15
Flaws	130	73	69
Tables without flaws	2	12	11
Tables needing rework	26	10	10

(a) Test to see if there are any differences in the rates at which flaws occur for the three evacuation times.

(b) Test to see if the percentages of tables with at least one flaw differ for the three evacuation times.

(c) Tables with more than two flaws require rework. Test to see if the percentages of tables that need rework are different for the three evacuation times.

(d) For each of the tests you performed, comment on its appropriateness. How much time should be spent evacuating the dry room?

3. A research firm conducted focus groups at five locations in Florida. The focus groups consisted of eco-tourists, theme-park tourists, beach-tourists, fishing-tourists, and nightlife-tourists. Before starting the focus group, each participant filled out a brief survey that included the question, "How much money per day have you been spending on accommodations?" The results are recorded below.

Dollars Spent on Accommodations				
Eco	Theme-park	Beach	Fishing	Nightlife
83	66	108	66	174
87	57	126	81	149
100	86	119	31	145
78	44	119	66	136
87	82	105	66	134
134	72	91	55	110

(a) Construct a normal probability plot to verify the appropriateness of using ANOM.

(b) Using ANOM, test to see if there is a difference in the average expenditures per day for accommodations for the five types of tourist.

4. To investigate child immunization rates in rural Nigeria, a researcher collected data from four government clinics and four church-sponsored clinics. Records for 80 randomly selected cases (subjects aged four to seven) were examined in each clinic. The

immunization rates for the four government clinics were 73.75%, 68.75%, 60.00%, and 66.25%. The four church-sponsored clinics had rates of 87.5%, 75.00%, 80.00%, and 67.50%.

   (a) Are there differences in immunization rates among the clinics?

   (b) Do the government clinics differ from the church-sponsored clinics with regard to immunization rate?

5. A researcher is interested in the effects of exercise on arthritic pain. Fifteen subjects with similar arthritis diagnoses were selected for a study in which five subjects got no exercise, five subjects got moderate exercise, and five subjects got strenuous exercise. Subsequent to the exercise, the time in minutes until a subject reports pain was measured. The data below are for day 1 of the experiment.

	Type of Exercise	
None	Moderate	Strenuous
44	11	96
4	77	3
0	6	164
0	16	54
25	79	32

   (a) Test to determine if the exercise level has any effect on the duration of the pain-free period.

   (b) Comment on the appropriateness of the test.

6. The experiment described in problem 5 was continued for 7 (exercise) days. The data below are the average durations of the pain-free periods for the 15 subjects.

	Type of Exercise	
None	Moderate	Strenuous
15.3	84.6	32.1
22.6	77.0	54.7
5.4	30.6	43.3
22.6	66.7	46.3
18.9	65.7	48.6

   (a) Test to determine if the exercise level has any effect on the duration of the pain-free period.

   (b) Comment on the appropriateness of the test, and compare your comments with those in problem 5.

7. A hospital administrator is interested in determining whether a health care plan affects the length of time (measured in weeks) patients receiving outpatient psychiatric treatment remain in treatment. Four insurance providers are to be compared. If the administrator wants to be 85% certain of detecting a difference of 1.5σ, how many patient files must be examined?

Chapter 3

One-Factor Unbalanced Studies

ANOM as described in the previous chapter assumes a *balanced design*. That is, it assumes that different factor levels all have the same number of observations. In practice, equal sample sizes often are not practical due to, for example, limitations on raw material supplies, data collection errors, or naturally unbalanced experiments. Observational studies are usually unbalanced. When a study does not have equal sample sizes, the ANOM test procedure is similar to the equal sample size procedure; however, it is slightly more complicated. The complication is due to the fact that the decision lines around $\bar{y}_{\bullet\bullet}$ will be tighter for the larger samples and wider for the smaller samples. As a result, different decision lines are needed for each sample size.

3.1 Normally Distributed Data

For normally distributed data, the decision lines for studies with unequal samples are (see Nelson (1989) for the mathematical details)

$$\bar{y}_{\bullet\bullet} \pm m(\alpha; I; N - I)\sqrt{\mathrm{MS}_e}\sqrt{\frac{N - n_i}{N n_i}} , \tag{3.1}$$

where n_i is the sample size for the ith treatment and the critical values $m(\alpha; I, \nu)$ are given in Table B.3. Similar to the $h(\alpha; I, \nu)$ values the $m(\alpha; I, \nu)$ critical values depend on

$$\alpha = \text{the level of significance desired,}$$
$$I = \text{the number of means being compared,}$$
$$\nu = \text{the degrees of freedom for MS}_e.$$

In addition, for unbalanced designs we need a more general formula for the MS_e, namely,

$$\mathrm{MS}_e = \frac{(n_1 - 1)s_1^2 + (n_2 - 1)s_2^2 + \cdots + (n_I - 1)s_I^2}{N - I} . \tag{3.2}$$

35

The assumptions for the ANOM are the same whether the sample sizes are balanced or unbalanced, namely, the ϵ_{ij} are approximately normally distributed and independent with approximately equal variances.

Example 3.1 (Blood Acid Data). As an initial step in demonstrating the validity of a new diagnostic test, the levels of a blood acid in healthy subjects were measured at three clinical sites. The sites did not enroll the same number of subjects within the study period, resulting in an *unbalanced* study. An initial study sought to determine if the acid levels of healthy subjects were the same at all three sites. If not, investigation into clinic differences would be needed before the remainder of the study could be performed. The normal probability plot of the residuals (i.e., the acid values centered by the clinic means) shown in Figure 3.1 indicates that the data are not normally distributed. To apply ANOM to this problem, one could transform the data by taking the natural logarithm of each value. The choice of this transformation is based on the right-skewed shape of the histogram (Figure 3.2) of the acid residuals. The normal probability plot of the residuals of the transformed data confirms that the transformation normalized the data (Figure 3.3). One can now proceed with ANOM applied to the transformed data. The summary statistics for the transformed data by clinic (i.e., site) are given in Table 3.1. To compare the three clinical sites, one would compute

$$\bar{y}_{\bullet\bullet} = \frac{75(5.46) + 74(5.34) + 105(5.34)}{75 + 74 + 105} = 5.375$$

and, using (3.2),

$$\text{MS}_e = \frac{75(0.73) + 74(0.73) + 105(0.53)}{254} = 0.647.$$

Using formula (3.1) and $\alpha = 0.05$, the three sets of ANOM decision lines around $\bar{y}_{\bullet\bullet}$ are

$$
\begin{aligned}
\text{Site 1:} \quad 5.375 \quad &\pm \quad m(0.05; 3, 251)\sqrt{0.647}\sqrt{\tfrac{254-75}{254(75)}} \\
&\pm \quad 2.39(0.0779) \\
&\pm \quad 0.186 \\
&\quad\; (5.19, 5.56);
\end{aligned}
$$

$$
\begin{aligned}
\text{Site 2:} \quad 5.375 \quad &\pm \quad 2.39\sqrt{0.647}\sqrt{\tfrac{254-74}{254(74)}} \\
&\pm \quad 0.188 \\
&\quad\; (5.19, 5.56);
\end{aligned}
$$

$$
\begin{aligned}
\text{Site 3:} \quad 5.375 \quad &\pm \quad 2.39\sqrt{0.647}\sqrt{\tfrac{254-105}{254(105)}} \\
&\pm \quad 0.144. \\
&\quad\; (5.23, 5.52).
\end{aligned}
$$

The ANOM chart is given in Figure 3.4, from which it is clear that there are no differences between clinics.

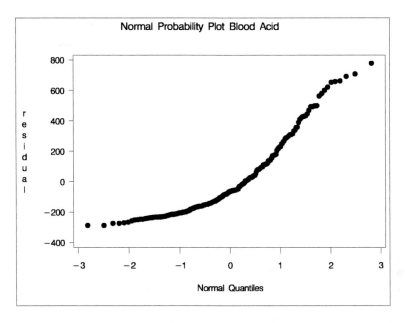

Figure 3.1. *Normal Probability Plot of the Blood Acid Data Residuals (Example 3.1).*

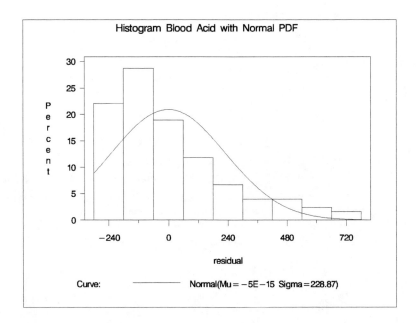

Figure 3.2. *Histogram of the Blood Acid Data Residuals (Example 3.1).*

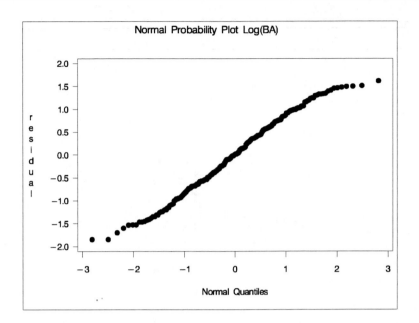

Figure 3.3. *Normal Probability Plot of the Transformed Blood Acid Data Residuals (Example* 3.1*).*

Table 3.1. *Transformed Blood Acid Data for Normal Subjects (Example* 3.1*).*

	Site	
1	2	3
$\bar{y}_{1\bullet} = 5.46$	$\bar{y}_{2\bullet} = 5.34$	$\bar{y}_{3\bullet} = 5.34$
$s_1^2 = 0.73$	$s_2^2 = 0.73$	$s_3^2 = 0.53$
$n_1 = 75$	$n_2 = 74$	$n_3 = 105$

Example 3.2 (Stream Remediation Data). An environmental consulting group collected five fish at each of three locations along a stream being monitored in a remediation program. The group was interested in the level of a heavy metal toxin in the fish. An equal number of fish were collected at each site; however, due to lab error, two samples from one site were destroyed, resulting in an unbalanced experiment. The data are given in Table 3.2. Now,

$$\bar{y}_{\bullet\bullet} = \frac{3(3.5) + 5(5.6) + 5(3.34)}{13} = 4.25$$

and, using (3.2),

$$\text{MS}_e = \frac{2(0.3)^2 + 4(0.19)^2 + 4(0.32)^2}{10} = 0.07.$$

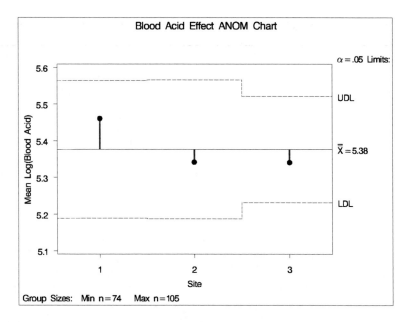

Figure 3.4. *ANOM Chart for the Transformed Blood Acid Data (Example 3.1).*

Table 3.2. *Toxin Level (g/kg).*

	Stream Site	
A	B	C
3.8	5.3	3.3
3.2	5.7	2.9
3.5	5.6	3.8
	5.8	3.3
	5.6	3.4
$\bar{y}_{A\bullet} = 3.5$	$\bar{y}_{B\bullet} = 5.6$	$\bar{y}_{C\bullet} = 3.34$
$s_A = 0.3$	$s_B = 0.19$	$s_C = 0.32$
$n_A = 3$	$n_B = 5$	$n_C = 5$

Also, $v = 13 - 3 = 10$ and $m(0.001; 3, 10) = 5.31$. So the decision lines are

$$
\begin{aligned}
\text{A:} \quad & 4.25 \quad \pm \quad 5.31\sqrt{0.07}\sqrt{\tfrac{13-3}{13(3)}} \\
& \quad\;\; \pm \quad 0.71 \\
& \qquad\qquad (3.54, 4.96);
\end{aligned}
$$

$$
\begin{aligned}
\text{B and C:} \quad & 4.25 \quad \pm \quad 5.31\sqrt{0.07}\sqrt{\tfrac{13-5}{13(5)}} \\
& \quad\;\; \pm \quad 0.94 \\
& \qquad\qquad (3.76, 4.74);
\end{aligned}
$$

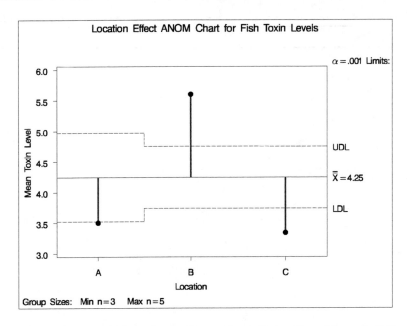

Figure 3.5. *ANOM Chart for the Stream Remediation Data (Example 3.2).*

the ANOM chart is given in Figure 3.5. From Figure 3.5 it is clear that there are significant differences (p-value ≤ 0.001) in the way the stream sites affect the toxin levels in the fish. Site B has significantly higher toxin levels, while sites A and C have significantly lower levels. Site B was a still area along the stream, while sites A and C had high water flows. This helps the consulting firm decide where to concentrate resources for further cleaning of the stream.

3.2 Binomial Data

When the proportions are based on unequal sample sizes one modifies the decision lines (3.1). The best estimator of $p(1 - p)$ is still $\overline{p}(1 - \overline{p})$, but now it is computed using

$$\overline{p} = \frac{n_1 \widehat{p}_1 + \cdots + n_I \widehat{p}_I}{N} .\qquad (3.3)$$

Note that $\overline{p}$ is really nothing more than the overall proportion of defectives in the combined samples. The df associated with this estimator are again ∞ since we are using the normal approximation. Taking all this into account results in the decision lines

$$\begin{array}{rcl} \text{UDL} & = & \overline{p} + m(\alpha; I, \infty)\sqrt{\overline{p}(1 - \overline{p})}\sqrt{\frac{N - n_i}{N n_i}} \\[2ex] \text{LDL} & = & \overline{p} - m(\alpha; I, \infty)\sqrt{\overline{p}(1 - \overline{p})}\sqrt{\frac{N - n_i}{N n_i}} \end{array} \qquad (3.4)$$

Again, before computing these decision lines, one would want to verify that the sample sizes are large enough to use the normal approximation by checking

$$n_i \widehat{p}_i > 5 \text{ and } n_i(1 - \widehat{p}_i) > 5 \text{ for } i = 1, \ldots, I . \qquad (3.5)$$

Example 3.3 (C-Section Data). A year-end review of a busy women's health clinic included looking at the proportion of c-section births performed by each of the four doctors in the clinic. The number of deliveries performed by each doctor differs, so the proportions that are of interest are based on unequal sample sizes. Are there any differences in the proportion of c-section deliveries at the 0.05 significance level?

The four sample proportions are

$$\widehat{p}_1 = \frac{64}{176} = 0.364, \quad \widehat{p}_2 = \frac{40}{137} = 0.292, \quad \widehat{p}_3 = \frac{8}{59} = 0.136, \quad \widehat{p}_4 = \frac{37}{160} = 0.231,$$

and (3.3) yields

$$\overline{p} = \frac{64 + 40 + 8 + 37}{176 + 137 + 59 + 160} = \frac{149}{532} = 0.280.$$

Checking condition (3.5), one would compute

$$\begin{array}{ll} n_1 \widehat{p}_1 = 64 & n_1(1 - \widehat{p}_1) = 112, \\ n_2 \widehat{p}_2 = 40 & n_2(1 - \widehat{p}_2) = 97, \\ n_3 \widehat{p}_3 = 8 & n_3(1 - \widehat{p}_3) = 51, \\ n_4 \widehat{p}_4 = 37 & n_4(1 - \widehat{p}_4) = 123. \end{array}$$

Since these quantities are all greater than 5, we can proceed with ANOM. For $\alpha = 0.05$, the decision lines (3.4) are

$$\text{Doctor 1:} \quad 0.280 \quad \pm \quad m(0.05; 4, \infty)\sqrt{0.280(1 - 0.280)}\sqrt{\tfrac{532-176}{532(176)}}$$
$$\pm \quad 2.49(0.0277)$$
$$\pm \quad 0.06897$$
$$(0.2110, 0.3490);$$

$$\text{Doctor 2:} \quad 0.280 \quad \pm \quad m(0.05; 4, \infty)\sqrt{0.280(1 - 0.280)}\sqrt{\tfrac{532-137}{532(137)}}$$
$$\pm \quad 2.49(0.0331)$$
$$\pm \quad 0.08242$$
$$(0.1976, 0.3624);$$

$$\text{Doctor 3:} \quad 0.280 \quad \pm \quad m(0.05; 4, \infty)\sqrt{0.280(1 - 0.280)}\sqrt{\tfrac{532-59}{532(59)}}$$
$$\pm \quad 2.49(0.0551)$$
$$\pm \quad 0.13720$$
$$(0.1428, 0.4172);$$

$$\text{Doctor 4:} \quad 0.280 \quad \pm \quad m(0.05; 4, \infty)\sqrt{0.280(1 - 0.280)}\sqrt{\tfrac{532-160}{532(160)}}$$
$$\pm \quad 2.49(0.0297)$$
$$\pm \quad 0.0740$$
$$(0.2060, 0.3540).$$

From the ANOM chart in Figure 3.6, one sees that there are significant differences in proportions due to doctor 1 delivering significantly more babies by c-section and doctor 3 delivering significantly fewer babies by c-section when compared to the overall clinic proportion. Review of patient records might be useful in determining why the proportion of c-sections was high for doctor 1.

Example 3.4 (Standardized Test Score Data), A school district has 563 fifth graders at 10 elementary schools of various sizes. The district measures school performance by students' standardized test scores. Is there a difference between schools based on the proportion of fifth graders scoring proficient on the standardized math test at an $\alpha = 0.01$ level? The data for the 10 schools (6 neighborhood schools and 4 alternative focus schools) are given in Table 3.3. The sample sizes are sufficiently large to apply ANOM. (One can check this by condition (3.5).) Using (3.3),

$$\overline{p} = \frac{50 + \cdots + 21}{52 + \cdots + 41} = \frac{415}{563} = 0.737.$$

Using $m(0.01; 10, \infty) = 3.29$ and (3.4), one would construct 10 pairs of decision limits, one for each school. The resulting ANOM decision chart in Figure 3.7 shows that three schools are performing at significantly low levels and two schools are performing at significantly high levels.

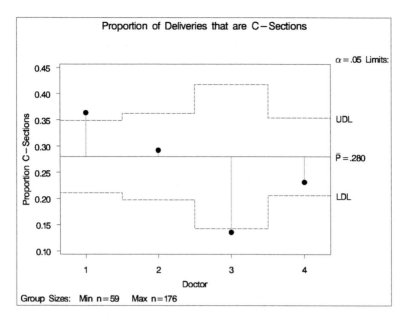

Figure 3.6. *ANOM Chart for the C-section Data (Example 3.3).*

Table 3.3. *Fifth Grade Math Proficiency Counts (Example 3.4).*

School	Enrollment	Proficient	$\widehat{p}_i$
N1	52	50	0.96
N2	55	15	0.27
N3	105	89	0.85
N4	81	75	0.93
N5	46	40	0.87
N6	84	67	0.80
Alt1	18	13	0.72
Alt2	33	28	0.85
Alt3	48	17	0.35
Alt4	41	21	0.51

3.3 Poisson Data

When the Poisson frequencies are based on different sample sizes, one must compute the average frequency as a weighted average using

$$\bar{u} = \frac{n_1\widehat{u}_1 + \cdots + n_I\widehat{u}_I}{N} \ . \tag{3.6}$$

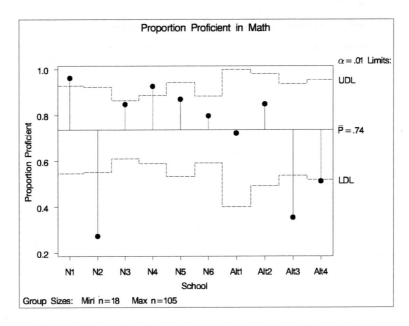

Figure 3.7. *ANOM Chart for the Test Score Data (Example* 3.4).

The decision lines (3.1) become

$$
\begin{array}{rcl}
\text{UDL} & = & \overline{u} + m(\alpha; I, \infty)\sqrt{\overline{u}}\sqrt{\frac{N-n_i}{Nn_i}} \\[2mm]
\text{LDL} & = & \overline{u} - m(\alpha; I, \infty)\sqrt{\overline{u}}\sqrt{\frac{N-n_i}{Nn_i}}
\end{array}
. \tag{3.7}
$$

Again, before computing these decision lines, one would want to verify that the sample sizes are large enough to use the normal approximation by checking that

$$
n\widehat{u}_i > 5 \quad \text{for } i = 1, \dots, I . \tag{3.8}
$$

Example 3.5 (Fiberglass Data). Four treatments were studied in the development of fiberglass with an end use in the molded fiberglass market. The choice of resin and the use (or nonuse) of a vacuum in the process were studied. Thirty-two test panels were molded for each treatment, and the number of defects in each panel was counted. Unfortunately, before all the panels could be evaluated, three of the panels produced with resin 2 and with the vacuum off were damaged. The numbers of undamaged panels and the numbers of defects are given in Table 3.4. To use ANOM to analyze this data, one needs

$$
\widehat{u}_1 = \frac{8}{32} = 0.25, \quad \widehat{u}_2 = \frac{6}{32} = 0.1875, \quad \widehat{u}_3 = \frac{11}{29} = 0.3793, \quad \widehat{u}_4 = \frac{10}{32} = 0.3125,
$$

Table 3.4. *Number of Defects for Undamaged Panels (Example 3.5).*

Resin	Vacuum	Defects	Number of Panels
1	Off	8	32
1	On	6	32
2	Off	11	29
2	On	10	32

and

$$\bar{u} = \frac{8 + 6 + 11 + 10}{32 + 32 + 29 + 32} = 0.28.$$

Checking the conditions (3.8), one obtains

$$32(0.25) = 25 > 5,$$
$$32(0.1875) = 6 > 5,$$
$$29(0.3793) = 11 > 5,$$
$$32(0.3125) = 10 > 5.$$

For $n_i = 32$ and $\alpha = 0.1$ the decision lines (3.7) are

$$\bar{u} \pm m(0.1; 4, \infty)\sqrt{\bar{u}}\sqrt{\frac{125 - 32}{125(32)}},$$

$$0.28 \pm 2.23(0.0807)$$
$$\pm 0.180$$
$$(0.10, 0.460),$$

and for $n_i = 29$ and $\alpha = 0.1$ the decision lines (3.7) are

$$\bar{u} \pm m(0.1; 4, \infty)\sqrt{\bar{u}}\sqrt{\frac{125 - 29}{125(29)}},$$

$$0.28 \pm 2.23(0.0861)$$
$$\pm 0.192$$
$$(0.088, 0.472).$$

The ANOM chart is given in Figure 3.8. Since none of the sample proportions fall outside the decision lines, there are no differences in the four treatments with respect to the defect rate.

Example 3.6 (CAT Scan Rate Data). A health care provider wishes to compare the number of CAT scans performed in the month of January at five clinics. Each clinic serves a different number of patients, so the health care provider would like to compare the number of CAT scans per clinic member. The number of CAT scans, the number of members assigned to

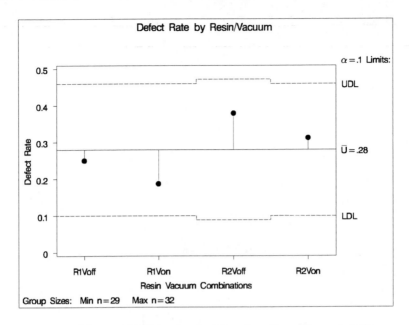

Figure 3.8. *ANOM Chart for the Fiberglass Data (Example* 3.5).

Table 3.5. *CAT Scans per Members Served (Example* 3.6).

Clinic	Number of Scans	Number of Members	Frequency ($\widehat{u}_i$)
A	50	26838	0.00186
B	71	26895	0.00264
C	41	26142	0.00157
D	62	25907	0.00239
E	89	26565	0.00335

each clinic, and the frequencies are given in Table 3.5. The Poisson distribution is applicable to this data since we are interested in a count per opportunity, where the count is the number of CAT scans and the opportunity is the number of members served by each clinic. Note that it is possible for a member to have more than one CAT scan in the time studied. To use the ANOM to look for significant differences in the number of CAT scans between clinics, we first check conditions 3.8. Doing so is equivalent to checking that the number of CAT scans is greater than five for each clinic. Using (3.6),

$$\bar{u} = \frac{50 + 71 + 41 + 62 + 89}{26838 + \cdots + 26565} = 0.00236,$$

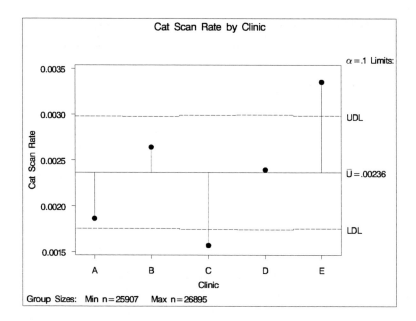

Figure 3.9. *ANOM Chart for the CAT Scan Data (Example 3.6).*

and for $n_1 = 26838$ and $\alpha = 0.10$ the decision lines (3.7) are

$$\bar{u} \pm m(0.1; 5, \infty)\sqrt{\bar{u}}\sqrt{\frac{132374 - 26838}{132347(26838)}}$$

$$0.00236 \pm 2.57(0.000265)$$
$$\pm 0.00068$$
$$(0.00168, 0.00304).$$

The ANOM chart in Figure 3.9 has decision lines for all five clinics, and one can see that there is one clinic with a below-average CAT scan rate and one clinic with an above-average CAT scan rate.

Chapter 3 Problems

1. To compare the effectiveness of three types of child immunization reminders, which are sent to parents a month before school starts, three large family practice groups were selected for study. Each of the three types of reminders was randomly assigned to a practice group. Families with children aged five through eight who had not been immunized were sent reminders. One month later data were collected at each practice group. The first type of reminder (a postcard) was sent to 81 families in practice group 1: 48 children received immunization and 14 had pending appointments to have immunizations. The second reminder (a postcard followed by a phone call) was

sent to 110 families in practice group 2: 55 children received immunization and 34 had pending appointments to have immunizations. The third reminder (a postcard offering a free kid's video to those who come for immunization) was sent to 67 families in practice group 3: 51 children received immunization and 9 had pending appointments to have immunizations. Are their differences in the effectiveness of the three types of reminder?

2. Three sunscreen additives thought to reduce the incidence of skin cancer are to be compared. Seventy-five volunteers with a history of precancerous facial lesions started the study. Twenty-five subjects were randomly assigned to each of the additives (identified as A, B, and C), and each month the subject was sent a free unmarked tube of ointment containing the appropriate additive. After 1 year, the subjects were examined for precancerous facial lesions. (The examiners were unaware of which additive each patient received.) The results are given in the table below. Note that not all the subjects completed the study. Are there differences in the effectiveness of the additives?

	Additive		
	A	B	C
Patients	25	19	15
Precancers	60	20	10
Patients without precancers	4	12	5

3. Refer to Problem 3 in Chapter 2. In practice, not all those scheduled to participate in a focus group arrive. Suppose that on average 75% arrive. To simulate this, for each data point perform an experiment with a 0.75 probability of success and delete those data points corresponding to failures. Repeat the data analysis with the new data set.

4. Refer to Problem 6 in Chapter 2. In studies of this type, subjects are allowed to drop out when they experience excessive pain.

   (a) Suppose from the moderate exercise group that subject 3 drops out and from the strenuous exercise group subjects 1 and 2 drop out. What effect does this have on the analysis and conclusions?

   (b) Suppose that each of these subjects dropped out after the third exercise session, but for reasons unrelated to pain, so that the data for these subjects represents their average pain-free period for three sessions. How might the analyst deal with this?

5. Small toy soldiers are made in an injection molding process requiring a front and a back mold. Ridge lines along the seam of the two molds have been a problem. These ridge lines are small and can appear in several places along the seam. One hour of production run time was allocated to test three new tube arrangements (the molds are fed through these tubes) to see whether this leads to a reduction in ridge lines. After allowing for setup time, 8 minutes of run time were expected to be available for each of the new arrangements and the standard one. The four arrangements were done in random order, resulting in the data below.

| | Arrangement | | | |
	NEW1	NEW2	NEW3	STANDARD
Toys produced	277	253	290	155
Ridges	42	48	70	17
Toys with ridges	20	10	45	15
Rework time (minutes)	26	29	57.5	16

The last run (STANDARD) was terminated after 5 minutes because setup times were longer than planned. Perform appropriate statistical tests to compare the arrangements.

Chapter 4

Testing for Equal Variances

The claim that all the populations under study have the same variance is called the homogeneity of variance (HOV) hypothesis. The HOV hypothesis for a single-factor experiment (study) with $k > 2$ factor levels can be represented as

$$H_0 : \sigma_1^2 = \cdots = \sigma_I^2, \tag{4.1}$$

where σ_i^2 is the variance of the ith population. The alternative hypothesis is not H_0.

ANOM, which is a test of location, can be converted into a test of scale (see Wludyka and Nelson (1997)) and used to test the HOV hypothesis. This test is called the analysis of means for variances (ANOMV) and is accomplished by transforming the observations y_{ij} into

$$t_{ij} = (y_{ij} - \overline{y}_{i\bullet})^2,$$

which when averaged become sample variances (with a multiplier). Power comparisons given by Wludyka and Nelson (1997) indicate that ANOMV is a good choice for testing hypothesis (9.1) when the assumption of normality is reasonable. In this chapter methods appropriate for balanced and unbalanced data will be presented, as well as small and large sample versions of ANOMV. For circumstances in which the normality assumption is inappropriate, two robust ANOMV tests will be presented.

4.1   ANOMV for Balanced Studies

Letting $\sigma^2 = \sigma_\bullet^2$ and $\tau_i = \sigma_i^2 - \sigma^2$, hypothesis (9.1) can be written as

$$H_0 : \tau_1 = \cdots = \tau_I = 0,$$

and estimates of the τ_i are

$$\widehat{\tau_i} = \overline{t}_{i\bullet} - \overline{t}_{\bullet\bullet}.$$

We start by assuming that there are samples of equal size n from each of the I populations. In that case,

$$\overline{t}_{i\bullet} = \frac{n-1}{n} s_i^2 \quad \text{and} \quad \overline{t}_{\bullet\bullet} = \frac{n-1}{n} \text{MS}_e$$

51

and, therefore,

$$\widehat{\tau_i} = \frac{n-1}{n}(s_i^2 - \mathrm{MS}_e).$$

Analogous to the ANOM test, one rejects hypothesis (9.1) if for some i the quantity $\widehat{\tau_i}$ (a measure of the discrepancy between the sample variance and the average of the I variances) is too large in magnitude. To quantify "too large" one needs to standardize the $\widehat{\tau_i}$ and find the joint distribution of these standardized variables under H_0. The standardized $\widehat{\tau_i}$ is (see Wludyka and Nelson (1997)) a linear function of

$$b_i = \frac{s_i^2}{\sum_{j=1}^{I} s_j^2}, \tag{4.2}$$

and thus for any decision rule based on the $\widehat{\tau}$ there is an equivalent decision rule based on the b_i.

Critical Values

Letting B_i represent the random variable describing the behavior of b_i, values L_α and U_α that satisfy

$$\Pr[L_\alpha \leq B_i \leq U_\alpha \text{ for } i = 1, \ldots, I] = 1 - \alpha \tag{4.3}$$

are lower and upper critical values for the B_i. Note that these critical values actually depend not only on α but also on the number of variances being compared (I) and the common df for the sample variances ($n-1$). They can be used to form the decision lines

$$\begin{array}{rcl}
\mathrm{UDL} & = & U(\alpha; I, n-1)I\mathrm{MS}_e \\
\mathrm{CL} & = & \mathrm{MS}_e \\
\mathrm{LDL} & = & L(\alpha; I, n-1)I\mathrm{MS}_e
\end{array}, \tag{4.4}$$

against which the sample variances s_i^2 are plotted. Hypothesis (9.1) is rejected if any s_i^2 falls outside the decision lines (4.4).

Conditions (4.3) do not provide unique critical values. To make the critical values unique, some additional criteria are necessary. A reasonable choice that is easy to implement is

$$\Pr(B_i > U_\alpha) = \Pr(B_i < L_\alpha) \text{ for } i = 1, \ldots, I. \tag{4.5}$$

For $I = 3$, critical values that satisfy (4.3) and (4.5) can be obtained via numerical integration and a search procedure. For larger values of I, numerical integration is not feasible. Instead one obtains bounds on L_α and U_α. These values are all given in Table B.4, where the conservative values have level of significance slightly smaller than α, and the liberal values have level of significance slightly larger than α.

Table 4.1. *Data and Summary Statistics (Example 4.1).*

	Brand		
1	2	3	4
87.7	83	90.9	86.5
100.4	87.2	94.2	71.4
85.9	86.7	85.6	80.7
105.5	86	98.5	94.4
74.9	80.4	99.7	97.8
75.3	86.3	108.5	97.9
$\bar{y}_{1\bullet} = 88.28$	$\bar{y}_{2\bullet} = 84.93$	$\bar{y}_{3\bullet} = 96.23$	$\bar{y}_{4\bullet} = 88.12$
$s_1^2 = 158.39$	$s_2^2 = 7.11$	$s_3^2 = 66.66$	$s_4^2 = 113.21$

Example 4.1 (Spring Data). The weight required to extend a spring 0.10 inches was determined for four different brands of springs, using six springs of each brand. The results are given in Table 4.1, together with the sample means and variances for each brand. Averaging the sample variances to obtain $MS_e = 85.589$ and using ANOMV with $\alpha = 0.1$, one obtains

$$UDL = U(0.1; 4, 5)IMS_e = (0.5863)(4)(85.589) = 200.72,$$

$$CL = MS_e = 85.59,$$

$$LDL = L(0.1; 4, 5)IMS_e = (0.0373)(4)(85.589) = 12.77.$$

From the ANOMV chart given in Figure 4.1, one sees that there is evidence of unequal variances at the 0.1 level because the variance of brand 2 is significantly small.

To evaluate practical significance, it is often more meaningful to look at standard deviations. An ANOMV decision chart can be constructed on which standard deviations are plotted. This is accomplished by taking the square roots of the lower decision line, center line, and upper decision line. This is illustrated in Figure 4.2, in which the standard deviation for brand 2 plots below the lower decision line, indicating significantly less variability for that brand.

Example 4.2 (Yogurt Data). The live culture count (count/10,000) was determined for three different types of starter using four batches of yogurt for each type starter. Of interest is the variability in the yogurt counts since the average can be targeted by adjusting the quantity of starter applied to the batch. The results are given in Table 4.2 with the sample means and variances for each brand. The quantities in parentheses are logs of the counts. The untransformed counts do not appear to be normally distributed, which is confirmed by examining the normal probability plot of the residuals in Figure 4.3. The log transformation of the data produces data that are reasonably normal (see Figure 4.4); hence ANOMV will be applied to the transformed data. Averaging the sample variances to obtain $MS_e = 0.088$

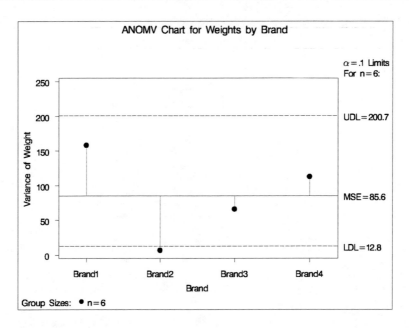

Figure 4.1. *ANOMV Chart for Tension Measurements of Four Brands of Springs (Example 4.1).*

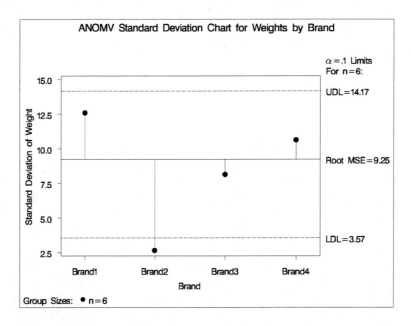

Figure 4.2. *ANOMV Standard Deviation Chart for Tension Measurements of Four Brands of Springs (Example 4.1).*

Table 4.2. *Data and Summary Statistics (Example 4.2).*

	Starter	
1	2	3
7.6(2.025)	2.5(0.922)	148.3(4.999)
4.5(1.513)	3.1(1.125)	299.5(5.702)
6.8(1.921)	2.5(0.934)	131.8(4.882)
6.9(1.926)	2.7(0.975)	106.9(4.672)
$\overline{Y}_{1\bullet} = 6.45(1.845)$	$\overline{Y}_{2\bullet} = 2.70(0.989)$	$\overline{Y}_{3\bullet} = 171.63(5.063)$
$s_1^2 = 1.817(0.054)$	$s_2^2 = 0.080(0.010)$	$s_3^2 = 7557.140(0.200)$

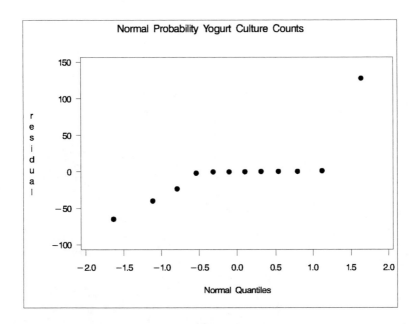

Figure 4.3. *Normal Probability Plot of Residuals of Yogurt Culture Counts for Three Starters (Example 4.2).*

and using ANOMV and with $\alpha = 0.1$, one obtains

$$\text{UDL} = U(0.1; 3, 3)I\text{MS}_e = (0.7868)(3)(0.088) = 0.2077,$$
$$\text{CL} = \text{MS}_e = 0.088,$$
$$\text{LDL} = L(0.1; 3, 3)I\text{MS}_e = (0.0277)(3)(0.088) = 0.0073.$$

From the ANOMV chart given in Figure 4.5, one sees that there is not sufficient evidence of unequal variances at the 0.1 level since all three variances plot within the decision lines. The three starters do not differ with respect to variability in counts.

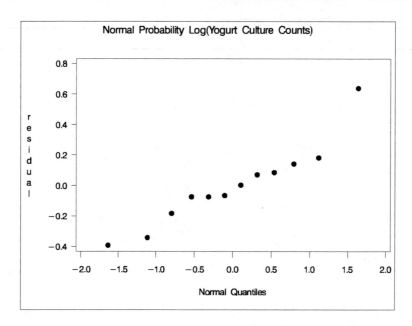

Figure 4.4. *Normal Probability Plot of Residuals Log of Yogurt Culture Counts for Three Starters (Example 4.2).*

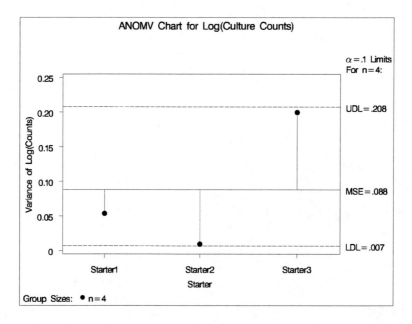

Figure 4.5. *ANOMV Chart for Log of Yogurt Culture Counts for Three Starters (Example 4.2).*

4.2 ANOMV with Unequal Sample Sizes

For the case of unequal sample sizes n_i, one has

$$\bar{t}_{i\bullet} = \frac{n_i - 1}{n_i} s_i^2 \quad \text{and} \quad \bar{t}_{\bullet\bullet} = \frac{N - I}{n_i I} \text{MS}_e$$

and, therefore,

$$\widehat{\tau}_i = \frac{n_i - 1}{n_i} s_i^2 - \frac{N - I}{n_i I} \text{MS}_e.$$

A standardized $\widehat{\tau}_i$ in this case is a linear function of

$$b_i^* = \frac{(n_i - 1)s_i^2}{(N - I)\text{MS}_e}.$$

Critical values in this case would depend on α, I, and all the n_i. For convenience when there is no confusion, we will suppress specific reference to these dependencies and use the notation

$$L_i^* = L^*(\alpha; I, \{n_i\}),$$
$$U_i^* = U^*(\alpha; I, \{n_i\}).$$

Letting B_i^* denote the random variable describing the behavior of b_i, the values L_i^* and U_i^* should satisfy

$$\Pr[L_i^* \leq B_i^* \leq U_i^* \text{ for } i = 1, \ldots, I] = 1 - \alpha,$$

and the decision lines are

$$\begin{aligned}
\text{UDL} &= U^*(\alpha; I, \{n_i\}) \left(\tfrac{N-I}{n_i-1}\right) \text{MS}_e \\
\text{CL} &= \text{MS}_e \\
\text{LDL} &= L^*(\alpha; I, \{n_i\}) \left(\tfrac{N-I}{n_i-1}\right) \text{MS}_e
\end{aligned}, \tag{4.6}$$

where MS_e is computed using (3.2).

Exact critical values are not available. If the n_i are not very different, one can use the conservative values in Table B.4 with df $= n_{\min} = \min_i\{n_i\}$. If the n_i are not close to each other, or if one wants slightly less conservative critical values, then one can use critical values that satisfy

$$\Pr(B_i^* > U_i^*) = \Pr(B_i^* < L_i^*) = \alpha/(2I). \tag{4.7}$$

Since $B_i^* \sim \text{Beta}(\theta_i, \sum_{j \neq i} \theta_j)$, where $\theta_i = (n_i - 1)/2$, this involves computing the inverse distribution function of a beta random variable (or searching for L_i^* and U_i^* if just the distribution function itself is available).

Example 4.3 (Velcro Force Data). Small kits containing blood sugar level test equipment are fastened together with velcro (two patches of velcro, one on the kit and the other on

Table 4.3. *Sample Sizes and Sample Variances for Velcro Force (Example 4.3).*

	Thickness								
	High			Medium			Low		
	Shape			Shape			Shape		
	C	D	S	C	D	S	C	D	S
n_i	5	5	5	8	8	8	14	14	14
s_i^2	115.18	40.37	59.78	30.60	173.56	80.43	49.27	3.34	9.50

Table 4.4. *Critical Values and Decision Lines (Example 4.3).*

n_i	θ_i	$\sum_{j \neq i} \theta_j$	L_i^*	U_i^*	UDL	LDL
5	2	34	0.00221	0.20918	191.45	2.03
8	3.5	32.5	0.01210	0.27605	144.37	6.33
14	6.5	29.5	0.04814	0.38862	109.44	13.56

the fastener latch). Since older people will be using these kits, it is important that there be little variation in the force required to pull open the fastener. Nine fastener designs consisting of combinations of three thicknesses of velcro (high, medium, low) and three shapes (circle, diamond, square) were used in an experiment. The quantity of velcro available in each thickness determined the sample sizes. These sample sizes and the resulting sample variances are given in Table 4.3. The critical values for $\alpha = 0.05$ were found using (4.7) with $\alpha/(2I) = 0.05/(2(9)) = 0.002778$. For example, for $n_i = 14$,

$$U_{n_i=14}^* = \text{Beta}^{-1}(1 - 0.002778, 6.5, 29.5) = 0.38862,$$

$$L_{n_i=14}^* = \text{Beta}^{-1}(0.002778, 6.5, 29.5) = 0.04814,$$

where the values for θ_i and θ_i, $\sum_{j \neq i} \theta_j$ are given in Table 4.4. (In EXCEL, e.g., one would use the function BETAINV.) Thus for $n_i = 14$, the decision lines are

$$\text{UDL} = 0.38862 \left(\tfrac{81-9}{14-1} \right) 50.85 = 109.44,$$

$$\text{CL} = 50.85,$$

$$\text{LDL} = 0.04814 \left(\tfrac{81-9}{14-1} \right) 50.58 = 13.56.$$

All the critical values and the resulting decision lines are given in Table 4.4, and the ANOMV chart is given in Figure 4.6. Since $s_{(M,D)}^2 = 173.56$ plots above its upper decision line and $s_{(L,D)}^2 = 3.34$ as well as $s_{(L,S)}^2 = 9.50$ plot below their lower decision lines, the hypothesis of equal variances is rejected. The combinations of low thickness with diamond shape and low thickness with square shape exhibit lower variability in the force required to open the kit.

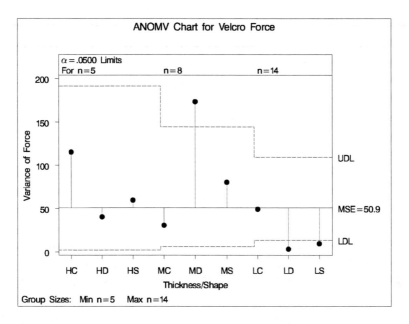

Figure 4.6. *ANOMV Chart for the Velcro Force Data (Example 4.3).*

4.3 ANOMV for Large Samples

For sample sizes greater than 35 (i.e., outside the range of Table B.4), one can use either

$$\Pr(B_i > U_i) = \Pr(B_i < L_i) = \alpha/(2I) \tag{4.8}$$

(or its counterpart for unequal sample sizes, equation (4.7)) or a large sample version of ANOMV. If the assumption of normality is reasonable, then using (4.8) or (4.7) will tend to result in tighter limits, particularly for the lower limit (see Examples 4.4 and 4.5). On the other hand, the large sample version does not rely on (or make use of) the normality of the underlying populations.

For independently and identically distributed observations from a population with finite kurtosis, the sample variance is asymptotically normal (see, e.g., Dudewicz and Mishra (1988, p. 326)). Thus, for large n_i the I sample variances s_i^2 can be thought of as playing the same role as the I sample means in ANOM.

Equal Sample Sizes

If the sample sizes are all equal, the ANOMV decision lines are

$$
\begin{aligned}
\text{UDL} &= \text{MS}_e + h(\alpha; I, \infty)\widehat{\sigma}_v \\
\text{CL} &= \text{MS}_e \\
\text{LDL} &= \text{MS}_e - h(\alpha; I, \infty)\widehat{\sigma}_v
\end{aligned}
\tag{4.9}
$$

where $\widehat{\sigma}_v$ is an estimate of the standard deviation of $s_i^2 - \mathrm{MS}_e$. When sampling from normally distributed populations,

$$\widehat{\sigma}_v = \mathrm{MS}_e \sqrt{\frac{2(I-1)}{I(n-1)}} \tag{4.10}$$

is an unbiased estimator. When the populations are clearly nonnormal, a reasonable estimate is

$$\widehat{\sigma}_v = \mathrm{MS}_e \sqrt{\frac{I-1}{I}\left(\frac{2}{n-1} + \frac{\widehat{\gamma}_2}{n}\right)}, \tag{4.11}$$

where $\widehat{\gamma}_2$ is an estimate of the common kurtosis.

Example 4.4 (Velcro Force Data). Since in Example 4.3 low thickness was associated with low variability, an additional 26 low-thickness fasteners of each shape were tested for a total of 40 observations with each shape. The resulting sample variances were 49.53 for the circular shape, 8.87 for the diamond shape, and 6.89 for the square shape so that

$$\mathrm{MS}_e = (49.53 + 8.87 + 6.89)/3 = 21.76$$

and (equation (4.10))

$$\widehat{\sigma}_v = (21.76)\sqrt{\frac{2(3-1)}{3(40-1)}} = 4.02.$$

Hence, the decision lines (4.9) with $\alpha = 0.01$ are

$$\begin{aligned}
\mathrm{UDL} &= 21.76 + h(0.01; 3, \infty)(4.02) = 21.76 + 2.91(4.02) = 33.45, \\
\mathrm{CL} &= 21.76, \\
\mathrm{LDL} &= 21.76 - h(0.01; 3, \infty)(4.02) = 21.76 - 2.91(4.02) = 10.06.
\end{aligned}$$

From the ANOMV chart given in Figure 4.7 it is clear that the diamond shape and the square shape have significantly lower variability, and the circular shape has significantly higher variability.

Example 4.5 (Velcro Force Data). Reanalyzing the data in Example 4.4 using (4.8), one finds $0.01/6 = 0.00167$,

$$\begin{aligned}
U(0.01; 3, 39) &= \mathrm{Beta}^{-1}(1 - 0.00167; 19.5, 39) = 0.5226, \\
L(0.01; 3, 39) &= \mathrm{Beta}^{-1}(0.00167; 19.5, 39) = 0.1722,
\end{aligned}$$

and the decision lines (4.4) are

$$\begin{aligned}
\mathrm{UDL} &= (0.5226)(3)(21.76) = 34.11, \\
\mathrm{CL} &= \mathrm{MS}_e = 21.76, \\
\mathrm{LDL} &= (0.1722)(3)(21.76) = 11.24.
\end{aligned}$$

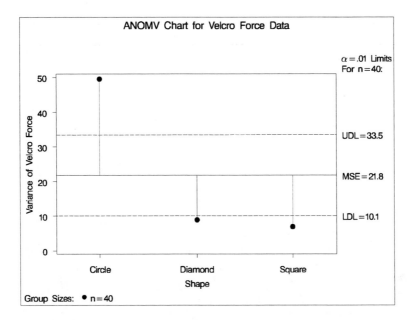

Figure 4.7. *ANOMV Chart for the Velcro Force Data (Example* 4.4).

Unequal Sample Sizes

When the sample sizes are all large, but are not equal, one can use the decision lines

$$
\begin{array}{rcl}
\text{UDL} & = & \text{MS}_e + m(\alpha; I, \infty)\widehat{\sigma}_{v_i} \\
\text{CL} & = & \text{MS}_e \\
\text{LDL} & = & \text{MS}_e - m(\alpha; I, \infty)\widehat{\sigma}_{v_i}
\end{array}
. \tag{4.12}
$$

Note that in this case the variance of $s_i - \text{MS}_e$ (and hence $\widehat{\sigma}_{v_i}$) depends on the sample size n_i. For normally distributed populations, an unbiased estimator is

$$
\widehat{\sigma}_{v_i} = \text{MS}_e \sqrt{\left(1 - \frac{n_i - 1}{N - I}\right)\left(\frac{2}{n_i - 1}\right)}. \tag{4.13}
$$

For obviously nonnormal populations when $\widehat{\gamma}_2 > 0$, a conservative estimate is

$$
\widehat{\sigma}_{v_i} = \text{MS}_e \sqrt{\left(1 - \frac{n_i - 1}{N - I}\right)\left(\frac{2}{n_i - 1} + \frac{\widehat{\gamma}_2}{n_i}\right)}, \tag{4.14}
$$

and when $\widehat{\gamma}_2 < 0$, a conservative estimate is

$$
\widehat{\sigma}_{v_i} = \text{MS}_e \sqrt{\left(1 - \frac{n_i - 1}{N - I}\right)\left(\frac{2}{n_i - 1} + \frac{\widehat{\gamma}_2}{n_i}\right) - \widehat{\gamma}_2 \frac{I - 1}{(N - I)^2}}. \tag{4.15}
$$

Table 4.5. *Summary Statistics for Electronic Components Lifetimes (Example 4.6).*

	Supplier			
	1	2	3	4
n_i	50	35	30	50
$\overline{y}_{i\bullet}$	34.61	35.51	33.23	28.62
s_i^2	24.57	234.76	59.84	269.11
$\widehat{\gamma}_{2_i}$	4.46	6.45	3.19	4.22

Example 4.6 (Electronic Component Lifetime Data). Each production line in a plant uses several dozen of the same electronic components. Since block replacement of the components is employed (all components on all lines are replaced at the same time), small variability in the lifetimes of the components is desirable. Components from each of four suppliers were tested and lifetimes were measured in tens of hours. The original design had called for testing 50 components from each supplier, but suppliers 2 and 3 were able to deliver only 35 and 30 components, respectively, in time for the test. The resulting summary statistics are given in Table 4.5. The sample kurtosis indicates that normality is not a reasonable assumption (since kurtosis for a normal population is zero). From the summary statistics one can compute

$$\text{MS}_e = \frac{(49)(24.57) + (34)(234.76) + (29)(59.84) + (49)(269.11)}{161} = 149.74,$$

$$\widehat{\gamma}_2 = \frac{(50)(4.46) + (35)(6.45) + (30)(3.19) + (50)(4.22)}{165} = 4.69.$$

For $n_i = 50$ one has (equation (4.14))

$$\widehat{\sigma}_{v_i} = 149.74 \sqrt{\left(1 - \frac{50 - 1}{165 - 4}\right)\left(\frac{2}{50 - 1} + \frac{4.69}{50}\right)} = 45.82,$$

and the $\alpha = 0.05$ decision lines (4.12) are

$$
\begin{aligned}
\text{UDL} &= 149.74 + m(0.05; 4, \infty)(45.73) = 149.74 + 2.49(45.82) = 263.8, \\
\text{CL} &= 149.74, \\
\text{LDL} &= 149.74 - m(0.05; 4, \infty)(45.82) = 149.74 - 2.49(45.82) = 35.65.
\end{aligned}
$$

Similar calculations for all the n_i are summarized in Table 4.6. From the ANOMV chart given in Figure 4.8, one finds that the variances are not all equal because the variance of supplier 1 is significantly smaller than average and the variance of supplier 4 is significantly larger than average.

Table 4.6. *Decision Lines for Electronic Components Lifetimes (Example 4.6).*

	Supplier			
	1	2	3	4
n_i	50	35	30	50
$\hat{\sigma}_{v_i}$	45.82	58.40	64.36	45.82
LDL_i	35.65	4.32	0	35.65
UDL_i	263.8	295.2	310.0	263.8

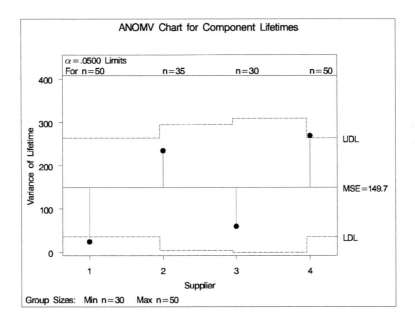

Figure 4.8. *ANOMV Chart for the Component Lifetimes (Example 4.6).*

4.4 Robust ANOM for Variance Test

When normality cannot be safely assumed, a robust test for variances is needed, since normality based tests such as ANOMV have unacceptably high type-I error rates for fat-tailed or skewed distributions (see, e.g., Wludyka and Nelson (1997)). Unlike ANOM and ANOVA, which exhibit large sample robustness under most practical situations, this problem is just as serious for large samples as small.

Example 4.7 (Bacterial Colony Data). Three bacterial growth inhibitors are to be compared. Each inhibitor is applied to 10 dishes. After 48 hours, each dish is examined and the number of bacterial colonies recorded. The data are given in Table 4.7. The bacterial colony counts are in the columns under Colonies. It is clear from inspection that these data are unlikely to have arisen by sampling from normal populations. A normal probability plot of the data for inhibitor I confirms this (see Figure 4.9). The plot suggests that the

Table 4.7. *Bacteria Colony Counts, Absolute Deviations from the Median (ADM) with Ranks in Parentheses, and Transformed Ranks for Three Samples of Size 10 (Example 4.7).*

	Colonies			ADM(Rank)			TR		
	I	II	III	I	II	III	I	II	III
	3	3	1	1.5 (16)	0.5 (7.5)	2 (18.5)	0.714	0.313	0.853
	5	2	0	0.5 (7.5)	0.5 (7.5)	3 (21)	0.313	0.313	1.012
	9	46	4	4.5 (23)	43.5 (28)	1 (13)	1.160	1.739	0.563
	1	17	3	3.5 (22)	14.5 (26)	0 (2)	1.083	1.448	0.082
	4	0	2	0.5 (7.5)	2.5 (20)	1 (13)	0.313	0.946	0.563
	5	2	3	0.5 (7.5)	0.5 (7.5)	0 (2)	0.313	0.313	0.082
	35	505	4	30.5 (27)	502.5 (30)	1 (13)	1.577	2.400	0.563
	4	1	13	0.5 (7.5)	1.5 (16)	10 (25)	0.313	0.714	1.340
	3	2	3	1.5 (16)	0.5 (7.5)	0 (2)	0.714	0.313	0.082
	99	8	5	94.5 (29)	5.5 (24)	2(18.5)	1.967	1.245	0.853
$\tilde{x}$	4.5	2.5	3.0						
$\bar{x}$	16.800	58.600	3.800	13.800	57.200	2.000	0.847	0.975	0.599
s^2	931.7	24797.4	12.6	888.2	24658.9	8.889	0.346	0.526	0.184

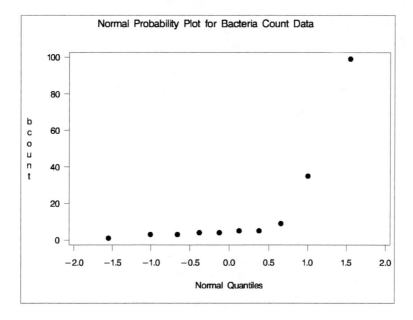

Figure 4.9. *Normal Probability Plot for Bacteria Counts for Treatment I (Example 4.7).*

population being sampled is right skewed. Hence, ANOMV would not be a good choice for testing the HOV hypothesis. Two approaches will be presented for dealing with this difficulty. (Recall that transforming the data to normality has been discussed, so that will not be considered here.) In the first approach, the original location measurements (the colony counts) are transformed into scale measurements to which ANOM can be applied. The second approach is to perform a randomization test, which is presented in Chapter 10.

Transformations

Two transformations that have been shown to produce robust HOV tests will be presented. In the first, the location measurements x_{ij} are transformed to absolute deviations from the median (ADM); that is,

$$y_{ij} = |x_{ij} - \tilde{x}_i| \tag{4.16}$$

for $i = 1, \ldots, I$ and $j = 1, \ldots, n_i$, where $\tilde{x}_i$ is the median of sample i. Note, for example, that the ADM for the first observation with inhibitor I is $y_{11} = |3 - 4.5| = 1.5$. The second transformation, denoted TR since it consists of transformations of the ranks of the ADMs, requires three steps. In step 1, the ADMs are calculated using (4.16). In step 2, one calculates $r_{ij} = \text{Rank}(d_{ij})$, in which the values from the combined sample are ranked from smallest to largest and the d_{ij} are the ADMs. Finally,

$$y_{ij} = \Phi^{-1}\left(0.5 + \frac{r_{ij}}{2N+1}\right), \tag{4.17}$$

where Φ^{-1} is the inverse normal score and N is the total number of observations. Hence, for example, the TR corresponding to the first observation with inhibitor I is $y_{11} = \Phi^{-1}(0.5 + 16/(60+1)) = \Phi^{-1}(0.762) = 0.714$. Note that the inverse normal score can be found using a statistical software package (such as SAS or EXCEL) or by lookup from a table of standard normal deviates (that is, $P(z < 0.714) = 0.762$).

ANOMV-LEV: The ANOM Version of Levene's Test

The HOV test by Levene (1960) is performed by transforming each observation using (4.16) and then applying the ANOVA F test to the resulting ADMs. In the ANOM version of Levene's test (ANOMV-LEV), the ANOM test is applied to the ADMs (4.16). Monte Carlo studies have shown both versions of Levene's test to be robust (see, e.g., Bernard and Wludyka (2001) as well as Conover et al. (1981)). In the balanced case one uses the decision lines defined in (2.4). In the unbalanced case, (3.1) is used.

Example 4.8 (Bacterial Colony Data). Recall the bacterial colony data from Example 4.7. The three ADM treatment averages (see Table 4.7) are $\overline{Y}_{1\bullet} = 13.8$, $\overline{Y}_{2\bullet} = 57.2$, and $\overline{Y}_{3\bullet} = 2.0$. The overall mean is $\overline{Y}_{\bullet\bullet} = 24.33$. The sample variances (also from Table 4.7) are averaged (equation (2.3)) to obtain

$$\text{MS}_e = \frac{888.2 + 24658.9 + 8.89}{3} = 8518.67$$

with $N - I = 30 - 3 = 27$ df. The critical value tables (Table B.1) are not indexed for df = 27. For $\alpha = 0.05$, one can construct conservative decision limits (i.e., they will be slightly wider than necessary) by using $h(0.05; 3, 24) = 2.5$. Alternatively, one could interpolate between the values $h(0.05; 3, 24) = 2.5$ and $h(0.05; 3, 30) = 2.47$ to obtain

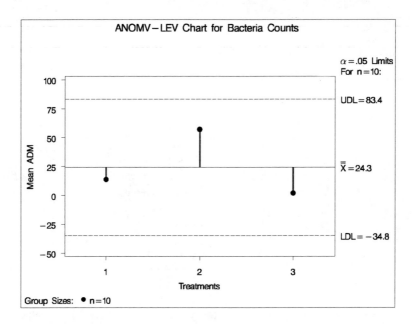

Figure 4.10. *ANOMV-LEV Chart for The Bacterial Colony Data (Example 4.8).*

$h(0.05; 3, 27) \approx 2.485$. Using the conservative value of 2.5,

$$h(0.05; 3, 24)\sqrt{\text{MS}_e}\sqrt{\frac{I-1}{N}} = 2.5\sqrt{8518.67}\sqrt{\frac{2}{30}} = 59.58.$$

Thus,

$$\text{UDL} = 24.33 + 59.58 = 83.91,$$
$$\text{LDL} = 24.33 - 59.58 = -35.25 = 0.$$

Note that since ADMs are greater than or equal to zero, the lower decision limit can be set to zero. Using a software package, one can use the exact df and obtain the limits shown in the ANOMV-LEV chart in Figure 4.10. (Note that the lower decision limit was not set to zero.) The decision lines make it clear that there is no difference in variability among the three bacterial inhibitors.

ANOMV-TR: The ANOM Version of the Fligner–Killeen Test

In the HOV test by Fligner and Killeen (as presented in Conover et al. (1981)), the ANOVA F test is applied to the transformed ranks of the absolute deviations from the median (4.17). ANOMV-TR, in which ANOM is applied to the transformed ranks of the ADMs, is described by Wludyka and Nelson (1999), who used a Monte Carlo study to demonstrate ANOMV-TR's robustness.

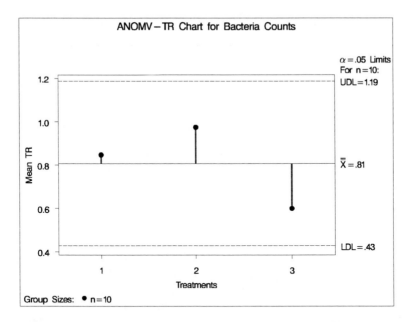

Figure 4.11. *ANOMV-TR Chart for The Bacterial Colony Data (Example 4.8).*

Example 4.9 (Bacterial Colony Data). To apply ANOMV-TR to the bacterial colony data, one needs to use the data under column TR in Table 4.7. It can be shown that (see problem 6) the $\alpha = 0.05$ decision lines for this example are

$$UDL = 1.185,$$
$$LDL = 0.428.$$

Similar to the conclusion reached using ANOMV-LEV, since the three TR means are within the decision limits (see Figure 4.11), one cannot conclude that the three bacterial inhibitors have different variability.

The data in this example are pseudorandom numbers in which each of the three populations are identical. Note that were one to inappropriately employ ANOMV in this case, one would conclude that the variances for the three inhibitors are different since the ANOMV decision limits for alpha = 0.05 are UDL = 128.9 and LDL = 48.64. (Using Table 4.7 it is evident that all three sample variances are outside the decision lines.)

It is worth noting that in general the ADMs will not be normally distributed (they are right skewed); nevertheless, ANOMV-LEV (similarly to Levene's ANOVA test) shows good robustness and can be safely employed in most situations.

The Odd Sample Size Case

For samples in which n_i is odd, there will be at least one ADM equal to zero for that sample. One can argue that since variability is being measured by the absolute deviation from the median, the median imparts no information about variability, and hence the zero ADMs should be deleted from the analysis. When this is done, the df will be reduced for each deleted zero ADM. Compensating for this, MS_e will on average be smaller (since for each sample the sample range is smaller). An example illustrating this idea can be found in Chapter 10.

4.5 Power and Sample Size Considerations for ANOMV

Power curves (see Wludyka et al. (2001)) have been constructed for balanced ANOMV. The power curves are indexed by α, the number of populations being compared, I, and

$$r = \frac{\sigma_{max}^2}{\sigma_{min}^2},$$

which is the ratio of the largest to the smallest variance. The latter is just a measure of the (projected) degree of dissimilarity among the variances. The curves plot the power (probability of rejecting the HOV hypothesis given r) versus the df $v = n - 1$. The power is given for the least favorable configuration of variances (see Wludyka et al. (2001) for details). To serve many practical problems, the power curves for $\alpha = 0.05$ and $\alpha = 0.10$ with $I = 3 - 8$, 10, and 12 variances have been summarized for power levels of 0.80 and 0.90 in Table B.5. The sample sizes in Table B.5 guarantee a minimum power of 0.80 or 0.90; for more exact powers one can consult the power curves found in Appendix A for $\alpha = 0.05$ and $I = 3 - 8$, 10, and 12 variances. Additional power curves can be found in Wludyka et al. (2002).

Example 4.10 (Coating Thickness). A quality engineer plans to conduct a study comparing the variability in coating thickness for three applicators. The coating will be applied to 1 inch^2 pieces and the thickness measured at the center of the piece. The engineer wishes to have probability 0.90 of being able to detect differences such that the largest standard deviation is three times the smallest standard deviation. She plans to use ANOMV with $\alpha = 0.05$. How many pieces should be tested for each applicator? In this case $I = 3$ and $r = 9$. Using the power curves in Figure A.9, for $v = 12$ the power is about 0.90; hence, a total of $3 \times (12 + 1) = 39$ pieces will have to be used. From Table B.5 for $\alpha = 0.05$, power $= 0.9$, the closest r value to 9 is $r = 10$, and for that one finds that 12 pieces are needed for each group. Interpolation between $r = 10$ and $r = 5$ would lead one to include 2 more pieces per group for a total of 39 pieces.

Example 4.11 A researcher was interested in variability in the effectiveness of six training methods. The researcher had enough volunteers to train five subjects with each method. At the end of the training a test was administered, producing a score for each subject. Using ANOMV with $\alpha = 0.05$, the HOV hypothesis was not rejected. A peer reviewer of the paper asked, What is the power of the test that was used? The largest sample standard deviation was 15.5 and the smallest was 3.6. One approach is to suppose that the observed values are similar to the actual variances, so that

$$r = \frac{\sigma_{max}^2}{\sigma_{min}^2} = \frac{15.5^2}{3.6^2} = 18.5.$$

The power, using Figure A.12 for $I = 6$ and $\alpha = 0.05$, is found where the curve for $\nu = 5 - 1 = 4$ intersects 18.5 on the x-axis, which occurs at about 0.30. This question could not be answered using Table B.5.

Comments for Practitioners Regarding ANOMV Tests

When appropriate, ANOMV should be used instead of ANOMV-LEV and ANOMV-TR. First, when normality is reasonable, ANOMV is more powerful. Second, the comparisons are made based on the variances (instead of the ADMs), which are more commonly used and better understood measures of variability. Plotting the standard deviations on a properly scaled decision chart often aids in assessing practical significance. When the assumption of normality is not reasonable, a transformation to normality can often be found. This choice takes advantage of the power of ANOMV; however, the original units of measurement have not been preserved, so interpretation and assessment of practical significance is not as direct.

When a robust test is necessary, in general ANOMV-TR is more robust than ANOMV-LEV (that is, it can be used in cases of extreme nonnormality); however, ANOMV-LEV is more powerful than ANOMV-TR. In most cases arising in practice, ANOM-LEV is robust enough and a better choice.

The reason for not using ANOMV for fat-tailed or skewed populations is the dangerously high Type I error rate. One approach to the problem (especially for large sample problems) is to apply ANOMV. If the HOV hypothesis in not rejected, then conclude that the variances are not different. If the HOV hypothesis is rejected, then examine the data for departures from normality. In cases in which the data is nonnormal because the tails are too short, simulation studies have shown ANOMV to be safe to use. If the data are skewed or fat-tailed, then verify your conclusions with a transformation or a robust ANOMV test.

Chapter 4 Problems

1. Refer to Problem 3 of Chapter 2. Test the hypothesis that the standard deviation in expenditures is the same for the five groups. What implication would rejection of the HOV hypothesis have regarding the ANOM test?

2. A health finance analyst wants to compare the variation in doctors' fees (for performing a c-section) in six suburban OB/GYN practices in Virginia. Seventy-five billing records are randomly selected for each practice. The data are summarized below. The units of measurement are thousands of dollars.

Practice	1	2	3	4	5
S_i	1.21	2.12	0.22	0.98	1.63

3. A state consumer protection officer is investigating auto repair facilities. The agency has subpoened the billing records from five brake repair shops and recorded the charges to the customers for all the front brake jobs done in the past week. The data are summarized below. The units of measurement are dollars. Test the hypothesis that the standard deviations are different for the five shops. Were one to reject the HOV hypothesis, what practical knowledge does that convey to the investigator?

Shop	1	2	3	4	5
n_i	17	9	23	14	11
S_i^2	126.11	488.41	1122.25	259.27	27.01

4. A lab manager is interested in assessing the skill of four technicians. A reagent of uniform concentration is divided into 20 batches of equal size. Each technician measures the concentration of five batches. The data are given in the table below. Precision and accuracy are both of interest. After verifying its appropriateness by means of normal probability plots, use ANOMV to test for equal variances among the technicians. Does this assess precision or accuracy? Suppose the concentration is actually 10 parts per milliliter. How can that fact be used to assess the skill of the technicians?

Batch	1	2	3	4
1	9.73	10.23	9.92	10.21
2	10.08	10.02	10.15	9.77
3	9.81	10.27	11.67	10.09
4	10.33	10.67	11.27	10.10
5	10.27	10.26	12.74	9.54

5. A process engineer wishes to investigate the variation in power in contact lenses (of the same nominal power) produced on three production lines. Ten lenses are selected from each production line and the deviation from target power recorded. The standard deviation of the deviation from target power for lines 1, 2, and 3, respectively, was 0.023, 0.092, and 1.2. It is known from experience that the deviation from target power is approximately normally distributed. Test that the standard deviation is the same for the three lines. Suppose that one lens from line 3 tore when it was removed from the package. How should the process engineer deal with this?

6. Compute the ANOMV-TR decision limits for the data in Example 4.7.

Chapter 5

Complete Multifactor Studies

To minimize the prediction variability to precisely determine what treatments in a study have significant effects, one must carefully plan how to collect the data, particularly when one is interested in studying more than one factor.

Suppose one is interested in studying the effects of time and temperature on the yield of a chemical process. The experimenter believes that whatever effect time has will be the same for the different temperatures. A common (but inefficient) approach is to study the effects of time and temperature separately. (This is referred to as a one-factor-at-a-time design.) Such a design might call for studying three times (for a fixed temperature) and taking three observations for each time. Then fixing time one might study two temperatures using three observations for each temperature. With these 15 total observations we can estimate both the average yield at a particular temperature and the average yield at a particular time. The uncertainty in the average yield for a fixed temperature would be less than that for a fixed time because more observations were taken (nine versus six).

A much more efficient way to study the two factors is to study them together. Taking two observations for each time-temperature combination would require only 12 total observations but would provide estimates for both the average yield at a particular temperature and the average yield at a particular time with uncertainties of only 87% ($= 100\%\sqrt{9/12}$) of the uncertainty for the average temperature from the previous design. Thus, by studying the two factors together, one can obtain better information with fewer observations.

The simplest experimental designs for studying two or more factors are complete layouts. Suppose one is interested in studying the effects of three drugs on hemoglobin levels (grams per deciliter) of male cancer patients who have undergone either chemotherapy or radiation therapy. I different therapies and J drugs are being considered. The I therapies are I levels of factor A, and the J drugs are J levels of factor B. Observations from this experiment could be arranged in an $I \times J$ table consisting of one row for each level of factor A and one column for each level of factor B, where the observations corresponding to the ith level of A and the jth level of B are recorded in cell (i, j). This is called a *two-way layout*, and if every cell in the table contains at least one observation, then it is a *complete layout*. Complete layouts are one example of a class of designs referred to as *factorial designs*. Data arising from observational studies can be analyzed as multiway layouts using

Table 5.1. *Data from a Factorial Design to Study Hemoglobin Levels in Males (Example 5.1).*

		Drug		
		Drug 1	Drug 2	Drug 3
		11.7	15.4	16.5
		13.6	14.0	15.9
	Chemo	14.5	12.9	15.4
		16.8	15.5	17.5
Therapy		11.7	12.6	16.1
		9.7	13.3	13.8
		12.3	11.8	12.7
	Radiation	12.3	11.8	12.9
		11.7	12.9	11.6
		13.1	9.1	14.6

the same methods and language arising from planned experiments. For an example driven summary of ANOM for two-way layouts, see Section 5.3.

Example 5.1 (Hemoglobin Data). The hemoglobin data given in Table 5.1 are a factorial design with $I = 2$, $J = 3$, and $n = 5$ observations per cell.

We will assume in this chapter that the multifactor experiments are complete and balanced. Incomplete designs are discussed in the next chapter.

5.1   Testing for Interaction

In addition to the efficiency obtained from the use of multifactor designs, one is also afforded the opportunity to study possible interaction between the factors. Two factors are said to interact if the relationship between the response and one factor depends on the level of the other factor. It is not possible to study interaction with one-factor-at-a-time experiments, and the unidentified presence of an interaction can lead to erroneous conclusions. With factorial designs, on the other hand, one can estimate interactions, and if none are present, the design allows for the effect of every factor to be evaluated as if the entire experiment were devoted entirely to that factor.

An interaction plot, which shows the effect of one factor for each level of a second factor, can be used to look for the possible presence of an interaction. For example, consider an extruded product (made from rubber and silica) that passes through a heated bath on a conveyor belt. The temperature of the bath and the speed of the belt can be controlled. An experiment was run to study the effect of the bath temperature and the belt speed on the end strength of the product. Consider the outcome illustrated by Figure 5.1, in which the temperature appears to have only a small effect (solid line with a slight upward slope) on strength when the flow rate is low. However, when the flow rate is fast, temperature appears to have a large effect on strength (dotted line with downward slope). If there were no interaction between temperature and flow rate, then the two lines would be parallel, as in Figure 5.2.

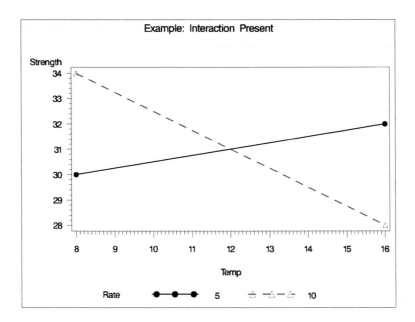

Figure 5.1. *An Interaction Plot Showing an Interaction.*

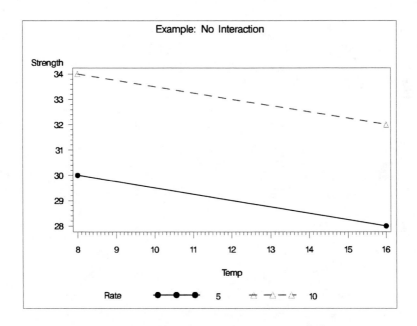

Figure 5.2. *An Interaction Plot Showing No Interaction.*

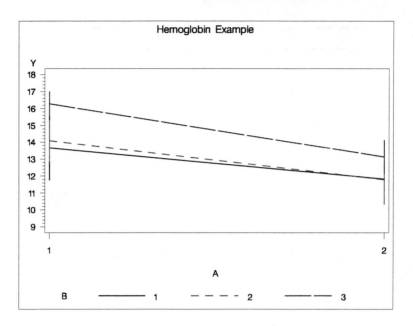

Figure 5.3. *An Interaction Plot for the Hemoglobin Data (Example 5.1).*

Whether the differences in the slopes of the line segments in an interaction plot are statistically significant depends on the experimental error. To indicate the magnitude of the experimental error, confidence intervals (i.e., error bars) are sometimes drawn around the end points of the line segments in an interaction plot. For example, the plot in Figure 5.3 is an interaction plot with error bars for the hemoglobin data (Example 5.1). However, since one is interested in comparing the slopes of the line segments, the confidence intervals on the end points provide only an indication regarding the presence of an interaction.

When there are $n > 1$ observations per cell, statistical tests for interactions can be performed. In the case of two factors, this is done using the model

$$\boxed{Y_{ijk} = \mu + \alpha_i^A + \alpha_j^B + \alpha_{ij}^{AB} + \epsilon_{ijk}}. \tag{5.1}$$

Notation.

$$\mu = \text{overall mean,}$$
$$\alpha_i^A = \text{main effect of the } i\text{th level of factor A,}$$
$$\alpha_j^B = \text{main effect of the } j\text{th level of factor B,}$$
$$\alpha_{ij}^{AB} = \text{interaction between the } i\text{th level of A and the } j\text{th level of B,}$$
$$\epsilon_{ijk} = \text{random variation of an individual experimental unit.}$$

As usual, we assume the ϵ_{ijk} are $N(0, \sigma^2)$ and independent. One could check the reasonableness of these assumptions using a normal probability plot of the residuals and the ANOMV test for equal variances. For model (5.1) the residuals are

$$\widehat{\epsilon}_{ijk} = y_{ijk} - \widehat{y}_{ijk}$$
$$= y_{ijk} - [\widehat{\mu} + \widehat{\alpha}_i^A + \widehat{\alpha}_j^B + \widehat{\alpha}_{ij}^{AB}],$$

where estimates for the parameters are

$$\widehat{\mu} = \overline{y}_{\bullet\bullet\bullet},$$
$$\widehat{\alpha}_i^A = \overline{y}_{i\bullet\bullet} - \overline{y}_{\bullet\bullet\bullet},$$
$$\widehat{\alpha}_j^B = \overline{y}_{\bullet j\bullet} - \overline{y}_{\bullet\bullet\bullet},$$

$$\widehat{\alpha}_{ij}^{AB} = \overline{y}_{ij\bullet} - \overline{y}_{i\bullet\bullet} - \overline{y}_{\bullet j\bullet} + \overline{y}_{\bullet\bullet\bullet} . \tag{5.2}$$

Combining all these equations, the residuals for model (5.1) reduce to

$$\widehat{\epsilon}_{ijk} = y_{ijk} - \overline{y}_{ij\bullet} . \tag{5.3}$$

That is, the residuals are calculated by subtracting the corresponding cell mean from each observation. With model (5.1), there are three general null hypotheses of interest:

$$H_A : \alpha_i^A = 0 \text{ for all } i,$$
$$H_B : \alpha_j^B = 0 \text{ for all } j,$$
$$H_{AB} : \alpha_{ij}^{AB} = 0 \text{ for all } (i, j).$$

These hypotheses correspond to the following: factor A has no effect; factor B has no effect; there is no AB interaction. The hypothesis H_{AB} must be tested first since the appropriate subsequent analyses depend on whether any interaction is present.

The ANOVA Test for Interaction

In general, the best way to test for interaction is to use ANOVA. This is done by combining the $\widehat{\alpha}_{ij}^{AB}$ terms to form the sum of squares for the AB interaction

$$SS_{AB} = n \sum_{i=1}^{I} \sum_{j=1}^{J} (\widehat{\alpha}_{ij}^{AB})^2 . \tag{5.4}$$

The df for SS_{AB} are the product of the df for factor A and the df for factor B, namely, $(I - 1)(J - 1)$. Therefore, the mean square for the AB interaction is

$$MS_{AB} = \frac{SS_{AB}}{(I - 1)(J - 1)} , \tag{5.5}$$

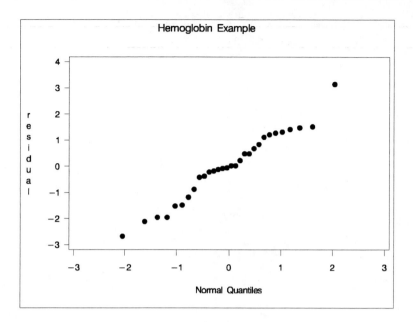

Figure 5.4. *Normal Probability Plot of the Residuals for the Hemoglobin Data (Example 5.2).*

and the test statistic for testing H_{AB} is

$$F_{AB} = \frac{\text{MS}_{AB}}{\text{MS}_e} \, , \tag{5.6}$$

where

$$\text{MS}_e = \frac{1}{IJ} \sum_{i,j} s_{ij}^2 \tag{5.7}$$

is simply the average of the sample variances for all the treatment combinations (the counterpart to (2.3)). The hypothesis H_{AB} is rejected if

$$F_{AB} > F(\alpha; (I-1)(J-1), IJ(n-1)).$$

Example 5.2 (Hemoglobin Data). To analyze the hemoglobin data in Example 5.1, one would first check the assumptions of normality and equal variances. A normal probability plot of the residuals is given in Figure 5.4, and there is no indication of lack of normality. To obtain the MS_e and test for equal variances, we need a table of the cell variances (Table 5.2). From the table of variances one finds

$$\text{MS}_e = \frac{4.563 + \cdots + 1.297}{6} = 2.110,$$

Table 5.2. *Cell Variances (Example 5.2).*

Therapy (A)	Drug (B)		
	Drug 1	Drug 2	Drug 3
Chemo	4.563	1.837	0.622
Radiation	1.652	2.687	1.297

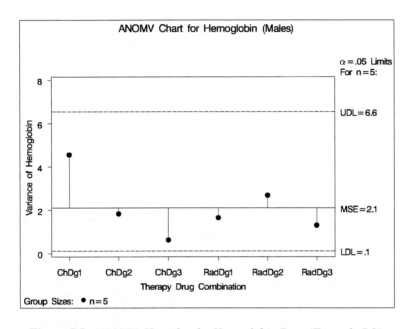

Figure 5.5. *ANOMV Chart for the Hemoglobin Data (Example 5.2).*

and the ANOMV decision limits are

$$\text{UDL} = U(0.05; 6, 4)I\text{MS}_e = (0.5175)(6)(2.110) = 6.55,$$
$$\text{CL} = \text{MS}_e = 2.110,$$
$$\text{LDL} = L(0.05; 6, 4)I\text{MS}_e = (0.0090)(6)(2.110) = 0.1139.$$

From the ANOMV chart in Figure 5.5, one sees that there is no evidence of unequal variances at the 0.1 level. The next step is to test for interaction between therapy and drug. It is easiest to compute the $\widehat{\alpha}_{ij}^{AB}$ by first computing a table of sample means (Table 5.3). From this table one would compute (using (5.2))

$$\widehat{\alpha}_{11}^{AB} = 13.66 - 14.67 - 12.74 + 13.46 = -0.297,$$
$$\widehat{\alpha}_{12}^{AB} = 14.08 - 14.67 - 12.93 + 13.46 = -0.067,$$
$$\widehat{\alpha}_{21}^{AB} = 11.82 - 12.24 - 12.74 + 13.46 = 0.297,$$

and so forth. All the $\widehat{\alpha}_{ij}^{AB}$ are given in Table 5.4. Using (5.4)–(5.6), one obtains

$$\text{SS}_{AB} = (5)[(-0.297)^2 + (-0.067)^2 + \cdots + (-0.363)^2] = 2.446,$$

$$\text{MS}_{AB} = \frac{2.446}{2} = 1.122,$$

$$F_{AB} = \frac{1.122}{2.110} = 0.532,$$

and since $0.532 < F(0.10, 2, 24) = 2.538$, there is no significant interaction between therapy and drug. This analysis is summarized in ANOVA table, Table 5.5.

Table 5.3. *Sample Means for the Hemoglobin Data (Example 5.2).*

Therapy (A)	Drug (B)			Therapy average
	Drug 1	Drug 2	Drug 3	
Chemo	13.66	14.08	16.28	14.67
Radiation	11.82	11.78	13.12	12.24
Drug Average	12.74	12.93	14.70	13.46

Table 5.4. $\widehat{\alpha}_{ij}^{AB}$ *(Example 5.2).*

Therapy (A)	Drug (B)		
	Drug 1	Drug 2	Drug 3
Chemo	−0.297	−0.067	0.363
Radiation	0.297	0.067	−0.363

Table 5.5. *ANOVA Table for the Hemoglobin Data (Example 5.2).*

Source	DF	Sum of Squares	Mean Square	F Value	Pr > F
Model	5	70.0216667	14.0043333	6.64	0.0005
Error	24	50.6320000	2.1096667		
Corrected Total	29	120.6536667			

R-Square	Coeff Var	Root MSE	Y Mean
0.580353	10.79368	1.452469	13.45667

Source	DF	Type I SS	Mean Square	F Value	Pr > F
A	1	44.40833333	44.40833333	21.05	0.0001
B	2	23.36866667	11.68433333	5.54	0.0105
A*B	2	2.24466667	1.12233333	0.53	0.5942

Table 5.6. *Results for the Process Yield Experiment (Example 5.3).*

	UV Level			
	UV		No UV	
Monomer 1	88.4	89.1	90.9	92.3
	90.9	90.8	92.3	92.3
Monomer 2	93.0	93.7	88.3	89.5
	93.4	92.9	88.0	89.6
Monomer 3	94.2	95.1	91.0	90.0
	96.4	95.7	92.4	92.8

Table 5.7. *Sample Means and Variances for the Process Yield Experiment (Example 5.3).*

Means						Variances			
		UV						UV	
		UV	No UV					UV	No UV
	1	89.80	91.95	90.88			1	1.553	0.490
Monomer	2	93.25	88.85	91.05		Monomer	2	0.137	0.670
	3	95.35	91.55	93.45			3	0.870	1.663
		92.80	90.78	91.79					

Example 5.3 (Process Yield Data). A 3×2 factorial experiment was designed to study the effect of three monomers and two levels of UV exposure (UV versus no UV) on the percent yield for contact lens production. Production runs were done four times with each monomer-UV combination (each production run was subjected to a complete automated inspection to determine the yield); results are given in Table 5.6. The 24 ($= 3 \times 2 \times 4$) production runs were made in random order (using a special research-and-development production line).

> **Note.** Each of the four replicates (i.e., the four production runs with each monomer-UV combination) required a complete run of the process. Simply taking four measurements from one run of the process with a particular monomer-UV combination would not suffice since that would not incorporate the run-to-run variability. Also, the trials were randomized (run in random order) to average out the effects of any unaccounted for factors.

From Table 5.6, a table of means (the average yield for each monomer-UV combination together with row means, column means, and the grand mean) and a table of sample variances were calculated (Table 5.7). Figure 5.6 shows a normal probability plot of the residuals and shows that the assumption of normality is reasonable. From the table of variances, one

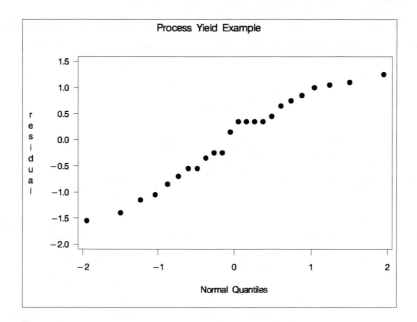

Figure 5.6. *Normal Probability Plot of the Residuals for the Process Yield Data (Example 5.3).*

can compute

$$MS_e = \frac{1.553 + \cdots + 1.663}{6} = 0.897$$

and the ANOMV decision lines

$$UDL = U(0.1; 6, 3)/MS_e = (0.5310)(6)(0.897) = 2.858,$$
$$CL = MS_e = 0.897,$$
$$LDL = L(0.1; 6, 3)/MS_e = (0.0066)(6)(0.897) = 0.036.$$

From the ANOMV chart in Figure 5.7, one sees that there is no evidence of unequal variances at the 0.1 level.

Using (5.2) and the table of means, one can compute the $\widehat{\alpha}_{ij}^{AB}$ in Table 5.8, and from that compute

$$SS_{AB} = 4(2)((-2.083)^2 + 1.192^2 + 0.892^2) = 52.443,$$
$$MS_{AB} = \frac{52.443}{2} = 26.222,$$
$$F_{AB} = \frac{26.222}{0.897} = 29.23.$$

Since $29.23 > F(0.05; 2, 18) = 3.555$, there is significant AB interaction. This confirms what one would have expected by looking at the interaction plot in Figure 5.8. Table 5.9 is the ANOVA output for this example.

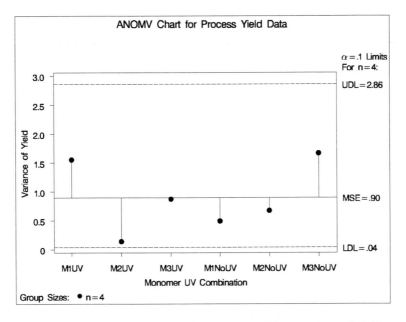

Figure 5.7. *ANOMV Chart for the Process Yield Data (Example 5.3).*

Table 5.8. $\widehat{\alpha}_{ij}^{AB}$ *for Example 5.3.*

		UV	
		UV	No UV
	1	1.371	−1.371
Monomer	2	−0.717	0.717
	3	−0.654	0.654

Using ANOM to Test for Interaction

One Factor at Two Levels

When one of the factors is at only two levels, there is an ANOM procedure to test for interactions. Suppose that $I = 2$ and $J \geq 2$, as in Example 5.1, where $I = 2$ and $J = 3$. A standard interaction plot in this situation would consist of J lines. (See the interaction plot for the data in Example 5.2) If the lines are parallel, then there is no interaction between factors; however, if the lines are clearly not parallel, then there is an interaction present. Therefore, checking for interaction amounts to comparing the slopes of these lines. The J slopes are

$$x_j = \overline{y}_{2j\bullet} - \overline{y}_{1j\bullet} \quad \text{for } j = 1, \dots, J,$$

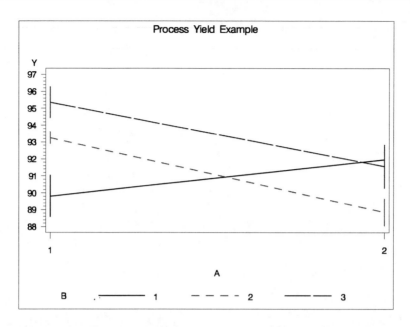

Figure 5.8. *Interaction Plot for the Process Yield Data (Example 5.3).*

Table 5.9. *ANOVA Table for the Process Yield Data (Example 5.3).*

Source	DF	Sum of Squares	Mean Square	F Value	Pr > F
Model	5	109.9683333	21.9936667	24.51	<.0001
Error	18	16.1500000	0.8972222		
Corrected Total	23	126.1183333			

R-Square	Coeff Var	Root MSE	Y Mean
0.871946	1.031922	0.947218	91.79167

Source	DF	Type I SS	Mean Square	F Value	Pr > F
A	1	24.40166667	24.40166667	27.20	<.0001
B	2	33.12333333	16.56166667	18.46	<.0001
A*B	2	52.44333333	26.22166667	29.23	<.0001

and these slopes can be compared using the ANOM decision lines (see Nelson (1988))

$$\overline{X}_{\bullet} \pm h(\alpha; J, v_{\epsilon})\sqrt{\mathrm{MS}_e}\sqrt{\frac{2(J-1)}{Jn}}, \tag{5.8}$$

where v_{ϵ} represents the df associated with MS_e.

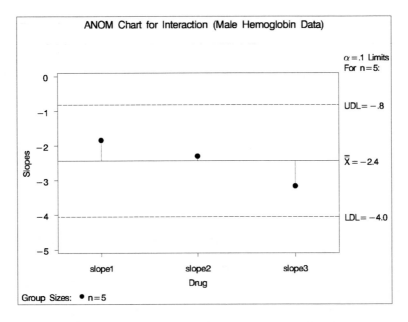

Figure 5.9. *ANOM Chart for Interaction (Example 5.2).*

Example 5.4 (Hemoglobin Data). For the hemoglobin data in Example 5.2, one can compute the three slopes from the table of means (Table 5.3) as

$$x_1 = 11.82 - 13.66 = -1.84,$$
$$x_2 = 11.78 - 14.08 = -2.30,$$
$$x_3 = 13.12 - 16.28 = -3.16,$$

and $\bar{x}_{\cdot} = -2.433$. The decision lines (5.8) are

$$-2.433 \pm h(0.10; 3, 24)\sqrt{2.110}\sqrt{\frac{4}{15}}$$
$$\pm 2.15(0.750)$$
$$\pm 1.61$$
$$(-4.04, -0.82),$$

and the ANOM chart is given in Figure 5.9, from which one sees that there is no significant interaction (i.e., one cannot reject the equal slopes hypothesis since each slope plots within the decision lines).

Example 5.5 (Process Yield Data). One can perform a similar analysis for the process yield data in Example 5.3. In that example, it is factor B (associated with j) that has two levels, so one must relabel the factors so UV is indexed by i and monomer is indexed by j. One

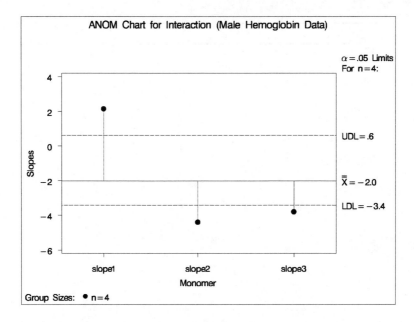

Figure 5.10. *ANOM Chart for Interaction (Example 5.3).*

would then compute the three slopes from the table of means as

$$x_1 = 91.95 - 89.80 = 2.15,$$
$$x_2 = 88.85 - 93.25 = -4.40,$$
$$x_3 = 91.55 - 95.35 = -3.80,$$

and $\overline{x}_\bullet = -2.017$. The decision lines (5.8) are

$$-2.017 \pm h(0.05; 3, 18)\sqrt{0.897}\sqrt{\frac{4}{12}}$$
$$\pm 2.55(0.546)$$
$$\pm 1.394$$
$$(-3.411, 0.623),$$

and the ANOM chart is given in Figure 5.10, from which one sees that there is a significant ($\alpha = 0.05$) interaction due to the slope of the line for monomer 1 being significantly greater than average, while the slopes for monomers 2 and 3 are significantly less than average. In addition, one can see that the slope for monomer 1 is positive, indicating that the direction of the UV effect is opposite when monomer 1 is used as compared to the negative slopes of monomers 2 and 3.

Both Factors at More Than Two Levels

Since the labeling of factors can be done in any order, we will assume that the labeling has been done such that $I \leq J$. When $I \geq 3$, the previous procedure can be generalized by

Table 5.10. *TG Level Reduction Data for Females (Example 5.6).*

Exercise	Intervention		
	No Intervention	Diet Management	Drug
Walking	−3.2	2.2	3.2
	0.3	−1.5	3.9
Bicycling	7.6	2.3	4.4
	3.0	4.7	−1.5
Running	5.5	6.0	24.3
	2.7	9.4	18.1

Table 5.11. $t_{j(ii')}$ *(Example 5.6).*

(i, i')	Intervention		
	No Intervention	Diet	Drug
(3,2)	−4.9	−0.7	8.0
(3,1)	1.9	2.5	5.9
(2,1)	3.1	−1.8	−13.9

comparing all possible pairings of the levels of factor A (which is associated with I). For each of the $\binom{I}{2}$ pairs of levels i and i', one would compute

$$x_{j(ii')} = \overline{y}_{ij\bullet} - \overline{y}_{i'j\bullet} \quad \text{for } j = 1, \ldots, J$$

and

$$t_{j(ii')} = x_{j(ii')} - \overline{x}_{\bullet(ii')} \quad \text{for } j = 1, \ldots, J.$$

The $t_{j(ii')}$ would then be plotted against the decision lines (see Nelson (1988))

$$0 \pm g(\alpha; (I, J), \nu_\epsilon)\sqrt{MS_e}\sqrt{\frac{2(J-1)}{Jn}}, \tag{5.9}$$

where ν_ϵ represents the df associated with MS_e; the critical values $g(\alpha; (I, J), \nu_\epsilon)$ are given in Table B.6.

Example 5.6 (Triglyceride Data). Consider a prospective randomized trial designed to determine factors affecting triglyceride (TG) levels in females 30 to 45 years old. Factor A is a type of exercise (walking, bicycling, running), and factor B is intervention (no intervention, diet management, cholesterol lowering drug). Hence, each factor is at three levels. The exercise consisted of 20 minutes of exercise three times per week. At enrollment in the study, each subject's TG level (mg/dL) was measured; after 120 days, a second TG measurement was made. The response variable is reduction in TG level. (Note that a similar baseline measurement could have been included in the hemoglobin study in Example 5.1 since baseline hemoglobin level is a likely covariate.) The results are given in Table 5.10, from which one can compute $MS_e = 8.111$. The $t_{j(ii')}$ are given in Table 5.11 and would

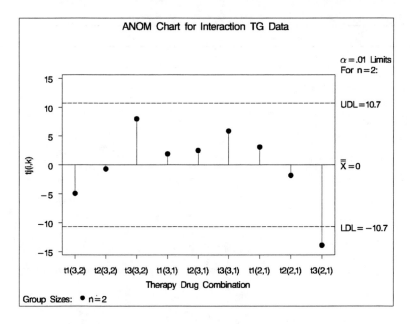

Figure 5.11. *ANOM Chart for Interaction (Example 5.6).*

be plotted against the decision lines

$$0 \pm g(0.01; (3, 3), 9)\sqrt{8.111}\sqrt{\frac{(2)(2)}{(3)(2)}}$$

$$\pm 4.60(2.325)$$

$$\pm 10.697.$$

From the decision chart in Figure 5.11 it is clear that there is a significant ($\alpha = 0.01$) interaction due to exercise type having a different effect when used in conjunction with the cholesterol lowering drug.

The ANOM procedure is a reasonable alternative to ANOVA for testing for interactions. For a small number of treatments, ANOM and ANOVA are essentially equally powerful (Nelson (1988)). In addition to testing for interaction, the ANOM output can be used to evaluate the nature of the interaction, although a basic interaction plot often illustrates this more clearly.

5.2   ANOM for a Two-Way Layout

When There Is No AB Interaction

When there is no significant AB interaction, one can study the effects of the two factors separately. Let f represent the number of levels for the factor of interest. That is, $f = I$

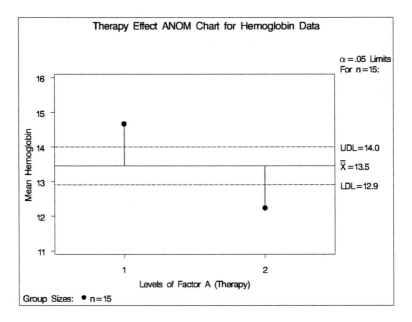

Figure 5.12. *ANOM Chart for Therapy (Example 5.7).*

for factor A and $f = J$ for factor B. The ANOM decision limits in this case are

$$\boxed{\bar{y}_{\bullet\bullet\bullet} \pm h(\alpha; f, IJ(n-1))\sqrt{\mathrm{MS}_e}\sqrt{\frac{f-1}{N}}} \quad . \tag{5.10}$$

Example 5.7 (Hemoglobin Data). Since in Example 5.1 we found no significant interaction between therapy and drug, we can study the two factors separately. For factor A (therapy), one obtains the ANOM decision lines

$$13.46 \pm h(0.05; 2, 24)\sqrt{\mathrm{MS}_e}\sqrt{\frac{1}{30}}$$

$$\pm 2.06\sqrt{2.110}\sqrt{\frac{1}{30}}$$

$$\pm 0.55$$

$$(12.91, 14.00)$$

and the resulting ANOM chart shown in Figure 5.12. Since the average hemoglobin level for radiation falls below the lower decision limit, and the average for chemotherapy falls above the upper decision limit, therapy has an ($\alpha = 0.05$) effect. Because there is a different number of levels for drug than for therapy, different ANOM decision lines need to be calculated to check for a drug effect. For factor B (drug), one obtains the ANOM

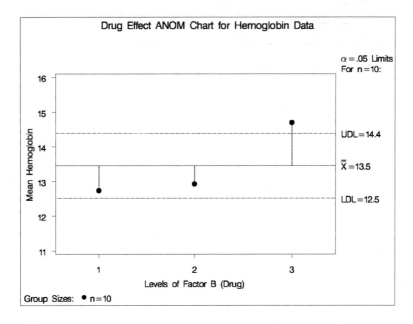

Figure 5.13. *ANOM Chart for Drug (Example 5.7).*

decision lines

$$13.46 \pm h(0.05; 3, 24)\sqrt{\mathrm{MS}_e}\sqrt{\frac{2}{30}}$$

$$\pm 2.50\sqrt{2.110}\sqrt{\frac{2}{30}}$$

$$\pm 0.94$$

$$(12.52, 14.40)$$

and the ANOM chart in Figure 5.13. There is an effect on hemoglobin level due to the drug since the average hemoglobin level for drug 3 is above the upper decision line.

When There Is an AB Interaction

If the ANOVA test (or the ANOM test) shows that there is significant interaction, one can't talk about the "main effect" of one factor or the other since the effect of one factor then depends on the particular level of the other factor. Instead, one approach is to reduce the data to one or more one-way layouts in which the problem of an interaction doesn't exist. One possibility is to split the data into subsets corresponding to the levels of one of the factors (generally the factor with fewer levels). If we split the data according to the levels of factor B, then we can study the effects of factor A separately for each level of factor B. At each level of factor B, one has a one-way layout with the treatments being the levels of

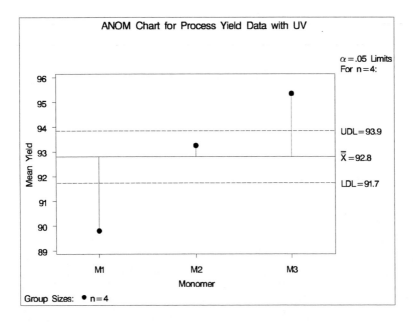

Figure 5.14. *ANOM Chart for Monomer with UV Exposure (Example 5.8).*

factor A. More specifically, for each level of factor B one would fit the model

$$Y_{ik} = \mu + \alpha_i^A + \epsilon_{ik}.$$

(5.11)

Note that this would require J separate analyses.

Example 5.8 (Process Yield Data). In Example 5.3 we found there was a significant interaction between the UV exposure and the monomer. Since there are only two UV exposure levels (but three monomers), it requires somewhat less work to consider the effect of the monomers separately for each UV exposure level using model (5.11). Using only the data for UV from Example 5.3, one obtains

$$\text{MS}_e = \frac{1.553 + 0.137 + 0.870}{3} = 0.853$$

and $\alpha = 0.1$ ANOM decision lines (equations (2.4))

$$92.80 \pm h(0.05; 3, 9)\sqrt{0.853}\sqrt{\frac{2}{12}}$$

$$\pm 2.79(0.3770)$$

$$\pm 1.052$$

$$(91.74, 93.85).$$

From the ANOM chart in Figure 5.14, one sees that the monomers are significantly different since monomer 1 produces significantly low average process yields and monomer 3 produces significantly high average process yields.

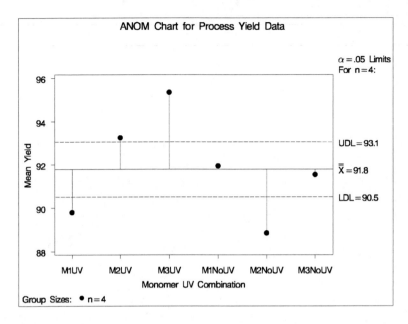

Figure 5.15. *ANOM Chart for the Monomer-UV Exposure Treatment Combinations (Example 5.9).*

A second possibility is to consider each of the $I \times J$ treatment combinations as one level of a single factor. That is, one could model the results of the experiment as

$$Y_{ijk} = \mu + \alpha_{ij}^A + \epsilon_{ijk}, \tag{5.12}$$

where α_{ij}^A is the main effect of the (i, j)th treatment combination. Which technique is more appropriate depends on the questions one is interested in answering. For the process yield data in Example 5.3, model (5.12) is probably more appropriate since the goal of studying the process would most likely be to find the UV exposure-monomer combination that results in the maximum process yield.

Example 5.9 (Process Yield Data). Analyzing the production process data in Example 5.3 using model (5.12), one would use $MS_e = 0.897$ from the analysis in Example 5.3 and compute ANOM decision lines (there are $3 \times 2 = 6$ treatment combinations)

$$91.79 \pm h(0.05; 6, 18)\sqrt{0.897}\sqrt{\frac{5}{24}}$$

$$\pm 2.91(0.432)$$

$$\pm 1.26$$

$$(90.53, 93.05).$$

From the ANOM chart in Figure 5.15, one sees that the treatment combinations are significantly different at the $\alpha = 0.05$ level due to the UV exposure with monomers 2 and 3 producing significantly high average yields (and two other combinations producing significantly low process yields). The ANOM decision chart puts statistical significance on what one observes in the basic interaction plot (Figure 5.8).

> **Note.** The above two approaches differ somewhat with respect to type I error control. In the second case, the level of significance is $\alpha = 0.05$. In the simple effects analysis, since each ANOM chart has level of significance $\alpha = 0.05$ and the tests are independent, the overall level of significance is $1 - (0.95)^2 = 0.0975$. If one wishes to maintain overall level of significance α for the simple effects analysis, then each of the type I ANOM charts should have (in the case where there are type I charts) level of significance $\alpha_i = 1 - (1 - \alpha)^{1/I}$. Using a software package such as SAS, this is direct, since any level of significance can be used; however, with lookup methods, one would have to use conservative limits from the tables.

With Only One Observation per Cell

The experiments we have looked at so far contain *replication*. That is, several experimental units were assigned to each treatment combination. Replication is always a good statistical practice. Without replication, we cannot be sure that the results we obtained from a particular set of conditions are due to the conditions rather than just random fluctuation. However, unreplicated experiments are often performed. Usually this is done to conserve resources, either when the total number of treatment combinations is large or when the experiment is either very expensive or time-consuming.

One of the complications of an unreplicated experiment is that we can no longer compute the MS_e by averaging sample variances. In an unreplicated experiment, we must find some other way to estimate σ^2. This is often done by assuming there is no interaction between the two factors, which results in the model

$$\boxed{Y_{ij} = \mu + \alpha_i^A + \alpha_j^B + \epsilon_{ij}} . \tag{5.13}$$

For model (5.13) the residuals are

$$\widehat{\epsilon}_{ij} = y_{ij} - \widehat{\alpha}_i^A - \widehat{\alpha}_j^B,$$

and substituting the estimators

$$\widehat{\alpha}_i^A = \overline{y}_{i\bullet} - \overline{y}_{\bullet\bullet},$$
$$\widehat{\alpha}_j^B = \overline{y}_{\bullet j} - \overline{y}_{\bullet\bullet},$$

one obtains

$$\widehat{\epsilon}_{ij} = y_{ij} - \overline{y}_{i\bullet} - \overline{y}_{\bullet j} + \overline{y}_{\bullet\bullet} . \tag{5.14}$$

Table 5.12. *Repair Cost Data.*

Material	5 mph	10 mph	15 mph	Average
		Speed(mph)		
plastic	529	979	1235	914.33
metal	851	1318	1852	1340.33
composite	474	897	1251	874.00
Average	618.00	1064.67	1446.00	1042.89

The residuals are then squared and summed to obtain

$$\text{SS}_e = \sum_{i,j} (\widehat{\epsilon}_{ij})^2 \tag{5.15}$$

and

$$\text{MS}_e = \frac{\text{SS}_e}{(I-1)(J-1)} . \tag{5.16}$$

The ANOM decision lines are similar to those in (5.10) except that the df in the critical value now reflect the df associated with the MS_e in (5.16), resulting in

$$\boxed{\overline{y}_{\bullet\bullet\bullet} \pm h(\alpha; f, (I-1)(J-1))\sqrt{\text{MS}_e}\sqrt{\frac{f-1}{N}}} . \tag{5.17}$$

Example 5.10 (Car Repair Data). An experiment was performed to understand how two factors (bumper material and speed) affected the cost of repairs due to low-speed front-end collisions. The data from the experiment are given in Table 5.12. The engineer did not anticipate any interaction between the two factors, and the interaction plot in Figure 5.16 shows no evidence of interaction (since the lines are nearly parallel). From (5.14) one obtains, for example,

$$\widehat{\epsilon}_{11} = 529 - 914.33 - 618.00 + 1042.89 = 39.56.$$

All the residuals are given in Table 5.13, from which one obtains

$$\text{SS}_e = (39.56)^2 + \cdots + (-26.11)^2 = 29387.11$$

and

$$\text{MS}_e = \frac{29387.11}{(2)(2)} = 7346.78.$$

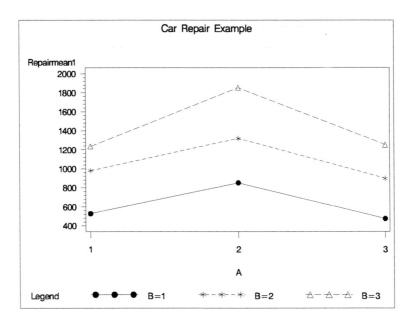

Figure 5.16. *Interaction Plot for the Car Repair Data (Example* 5.10*).*

Table 5.13. *Residuals for the Car Repair Data (Example* 5.10*).*

Material	Speed (mph)		
	5	10	15
plastic	39.56	42.89	−82.44
metal	−64.44	−44.11	108.56
composite	24.89	1.22	−26.11

Since interaction is not an issue (we have assumed it doesn't exist), we can analyze the effects of speed and material using the ANOM. With $\alpha = 0.01$, the decision limits for speed are (formula (5.17))

$$1042.89 \pm h(0.01; 3, 4)\sqrt{7346.78}\sqrt{\frac{2}{9}}$$

$$\pm 5.74(40.406)$$

$$\pm 231.93$$

$$(810.96, 1274.82).$$

Since speed and material have the same number of levels, their decision limits would be the same. The two ANOM charts are given in Figures 5.17 and 5.18, from which one sees that speed is significant at the $\alpha = 0.01$ level because the 5 mph rate is significantly low and the 15 mph rate is significantly high. The material is significant at the $\alpha = 0.01$ level because the repair cost for metal is significantly high.

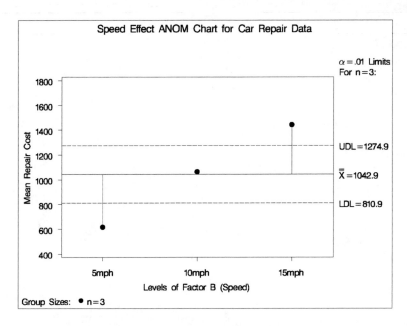

Figure 5.17. *ANOM Chart for Speed (Example 5.10).*

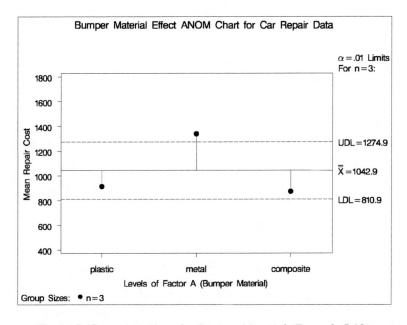

Figure 5.18. *ANOM Chart for Bumper Material (Example 5.10).*

Randomized Block Designs

To decrease the variability associated with the factors of interest in an experiment, the experimental trials can sometimes be split into subgroups in which the experimental conditions are more homogeneous than they are over the entire experiment. These subgroups are called blocks, and a factor used to accomplish this subdivision is called a blocking factor.

The simplest example of blocking is a paired experiment. For example, suppose one is interested in comparing the durability of two UV coatings for eyeglasses. Failure to account for differences between the test subjects (the blocks) would contaminate the experiment. Two different consequences are possible, depending on exactly how the experiment was conducted, and both are bad. If one UV coating was used on glasses worn by test subjects who were meticulous about cleaning and caring for their lenses, and the other type was used on glasses worn by subjects who took much less care of their lenses, one would not be able to distinguish between differences due to the UV coating and differences due to its use. This is a particularly bad design since one would at the least want to randomize over any factor (in this case the subjects) that might affect the outcome. Alternatively, if one randomized over the subjects, that is, assign at random the type of coating to the subject (this would be referred to as a *completely randomized design*), then the additional variability due to the differences in the way the subjects use their glasses would make it more difficult (or maybe impossible) to find differences in the UV coatings when they really existed.

The appropriate design in this case is called a *randomized block design*. This is a design in which the different experimental conditions are accounted for by assigning each level of the factor of interest to each block, and additional randomization is done within the blocks. In the lens example, this would amount to providing each subject with one lens of each type, and the randomization would consist of assigning the two UV coating types to the left and right eyes at random.

Other examples of blocking factors are batches of raw material, technicians, and days on which experimental trials are performed. In each instance, the experimenter is not specifically interested in the effect of the blocking factor but in separating the variability associated with it from the variability associated with the factors of interest. It is possible to have more than one blocking factor in an experiment, such as the situation in which experimental trials are performed by different technicians over several days. The analysis associated with a blocking factor is the same as for any other factor. The only difference is that because of their nature, blocking factors are assumed not to interact with the other factors. The blocking factor can be a random effect (not a fixed effect), so some practitioners will employ a mixed model in their analysis. Here the primary focus is on analyzing the main effects associated with the fixed effect factor.

Example 5.11 (Dye Irritation Data). Three dyes can be used to create tattoos. To aid in the choice of dye that minimizes post-tattoo irritation, researchers conduct a small experiment in which five subjects (volunteers) are to be marked with each of the dyes. (Three small rectangular regions are identified on the left forearm of each subject.) The assignment of dye to location on each arm is decided randomly. Each marking is then observed until the adjacent region of irritation is less than 1 cm. The time (hours) until the irritation is sufficiently reduced is recorded (lower numbers are better). The data are recorded in Table 5.14. This is a randomized block design in which the five subjects (volunteers) are the blocks, and the dyes are the treatments (the levels of the factor of interest). Table 5.15

Table 5.14. *Dye Irritation Data (Example 5.11).*

	Subject					Average
Dye	1	2	3	4	5	
1	64	68	66	72	67	67.4
2	71	74	79	81	62	73.4
3	85	87	81	76	81	82.0
Average	73.3	76.3	75.3	76.3	70.0	74.3

Table 5.15. *Residuals for the Dye Irritation Data (Example 5.11).*

	Subjects				
Dye	1	2	3	4	5
1	−2.47	−1.47	−2.47	2.53	3.87
2	−1.47	−1.47	4.53	5.53	−7.13
3	3.93	2.93	−2.07	−8.07	3.27

contains the residuals (equation (5.14)). From (5.15) and (5.16) one obtains

$$SS_e = [(-2.47)^2 + \cdots + (3.27)^2] = 246.13$$

and

$$MS_e = \frac{246.13}{8} = 30.77.$$

Using $\alpha = 0.05$, the ANOM decision lines (equation (5.17)) are

$$74.27 \pm h(0.05; 3, 8)\sqrt{30.77}\sqrt{\frac{2}{15}}$$

$$\pm 2.86(2.026)$$

$$\pm 5.79$$

$$(68.48, 80.06),$$

and the ANOM chart is shown in Figure 5.19. From the ANOM chart, one sees that the dyes are significantly different ($\alpha = 0.05$) because dye 1 is significantly good and dye 3 is significantly bad.

5.3   Practitioner's Summary of Two-Way ANOM

Suppose that one has data arising from a two-factor study, factor A at I levels and factor B at J levels, usually identified as an $I \times J$ full factorial with n observations at each factor level combination. This section provides steps for ANOM analysis for balanced data of this type. It is important to recall that the assumptions for ANOM are identical with the assumptions for ANOVA. Furthermore, the underlying data models are identical, so any one

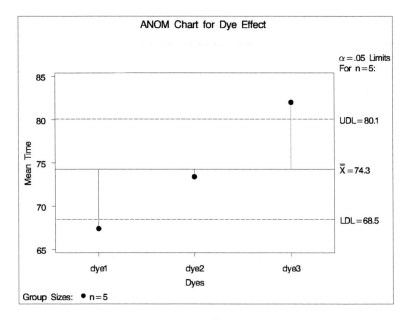

Figure 5.19. *ANOM Chart for the Dye Irritation Data (Example 5.11).*

of many statistical software packages can be used to do parts of the analysis. Software with an ANOM procedure can of course be used to directly create the ANOM decision charts. For this analysis, we will think of the data as consisting of an $I \times J$ table.

The key idea is that the analysis of two-way data will produce one or more one-way ANOM decision charts. The following steps summarize the analysis:

1 Calculate cell, row, column, and overall means.

2 Investigate AB interaction using a means interaction plot. Usually a visual examination will reveal the extent and nature of potential interaction. For $n > 1$, ANOVA can be used to test for AB interaction by using the full effects model (5.1) and performing the F test provided in a software package. If interaction is judged to be important, go to step 3A; otherwise go to step 3B.

3A When significant interaction is present, practitioners may use one of three approaches: (1) a simple effects analysis, in which ANOM decision charts are constructed for each level of one factor; this results in either I or J ANOM decision charts; (2) removal of the structure by consolidating the data into a one-way layout with $I \times J$ levels and constructing one ANOM decision chart; or (3) a compromise in which levels of either factor A or factor B are combined and the simple effects analysis is applied to this structure. Note that MS_e can be found in the standard ANOVA table outputted by software packages. For approach (1), an ANOVA table is produced for each level of the factor held constant, and the ANOM chart is based on the corresponding MS_e from the ANOVA table. For approach (2), MS_e can be found in the single ANOVA table associated with the one-way layout.

Table 5.16. *ANOVA Table for the TG Data (Example 5.12).*

Source	DF	Sum of Squares	Mean Square	F Value	Pr > F
Model	8	701.5111111	87.6888889	10.81	0.0009
Error	9	73.0000000	8.1111111		
Corrected Total	17	774.5111111			

R-Square	Coeff Var	Root MSE	Y Mean
0.905747	56.08755	2.848001	5.077778

Source	DF	Type I SS	Mean Square	F Value	Pr > F
A	2	335.9344444	167.9672222	20.71	0.0004
B	2	124.5877778	62.2938889	7.68	0.0113
A*B	4	240.9888889	60.2472222	7.43	0.0063

3B When significant interaction is not present, practitioners can perform two ANOM analyses, one for each of the main effects. For the case in which $n > 1$, MS_e can be found in the ANOVA table for the full effects model (the one used to investigate interaction). In the case in which $n = 1$ MS_e is in the main effects model (5.13) ANOVA table (no AB interaction term).

4 Since the assumptions and models for ANOM are identical to those in ANOVA, analysis of the residuals would proceed exactly as in ANOVA. Most software packages calculate the residuals from which one can test for normality with a normal probability plot or a numerical statistical tests for normality. For the investigation of the HOV assumption, one can look at comparative box plots, soft comparison of variances of the cell residuals (or, equivalently, the cell variances of the data), or use a formal HOV tests, such as ANOMV (see Chapter 4). Plots can be used to determine whether the independence assumption is reasonable.

Some practitioners may choose to handle step 4 sooner. In particular, if the data require transformation to achieve normality, one may wish to establish that early in the analysis. The steps above will be illustrated by displaying output from SAS 9.1.2. (See Appendix C for SAS examples.) Apart from the ability to generate the ANOM decision charts directly, the choice of software doesn't matter. Three examples will be presented: one with interaction, one without interaction, and one with only one observation per cell.

Example 5.12 (TG Data). A complete ANOM analysis of the TG data (Table 5.10) from Example 5.6 will be presented. Recall that this is a 3×3 study with 2 replicates per cell, so ANOVA (based on the model with interaction) may be used to determine whether significant interaction exists. From Table 5.16 it is evident that significant exercise-intervention interaction is present ($p = 0.0063$). The nature of the interaction is revealed in the means plot (see Figure 5.20). We will first do a simple effects analysis, examining the effect of intervention (factor B) for each level (type) of exercise (factor A). This analysis makes sense if one assumes that persons will choose an exercise type and then one wishes to determine the

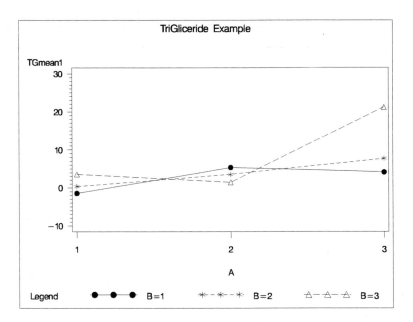

Figure 5.20. *Standard Means Interaction Plot for the TG Data (Example 5.12).*

best intervention for that exercise type. This will result in three ANOM charts, one for each level of the exercise factor. Using a software package with ANOM capability, one can do this directly, from which one can see that for walking (Figure 5.21) and biking (Figure 5.22) there are no significant differences among the interventions (since in each case all factor level means plot within the decision lines); however, treatment with the cholesterol lowering drug (level 3 of factor B) produces a significant reduction in TG for subjects engaged in running (Figure 5.23).

For packages without ANOM capability, charts can easily be constructed using ANOVA output. We will illustrate the idea by analyzing exercise level 3 (running). The ANOVA output, from which one obtains $MS_e = 9.64$ and df $= 3$, is in Table 5.17. For convenience, the factor levels means for B have been included. Choosing $\alpha = 0.05$, the ANOM decision lines (2.4) are

$$11.00 \pm h(0.05; 3, 3)\sqrt{9.64}\sqrt{\frac{2}{6}}$$
$$\pm 4.18(1.79)$$
$$\pm 7.49$$
$$(3.51, 18.49).$$

Using these decision limits and the factor level means in Table 5.17, one can manually create a chart similar to the one in Figure 5.23.

An alternative is to treat the problem as a one-way layout, which makes sense if one is interested in finding the best combination of exercise and intervention. Treating the problem

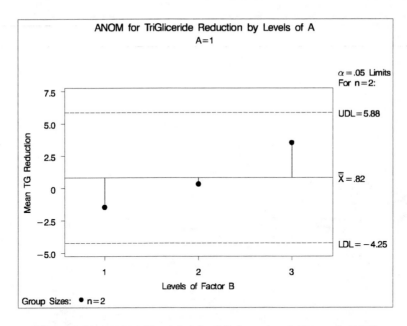

Figure 5.21. *ANOM Chart for the TG data; A = 1 (Example 5.12).*

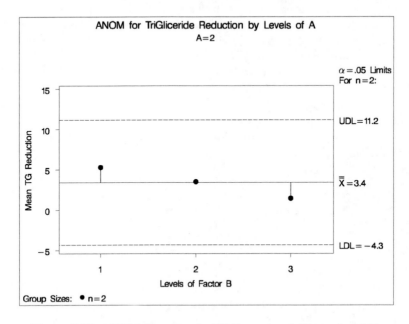

Figure 5.22. *ANOM Chart for the TG Data; A = 2 (Example 5.12).*

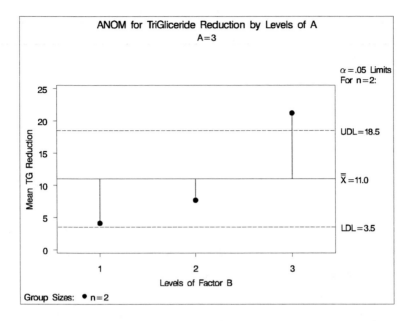

Figure 5.23. *ANOM Chart for the TG Data; A = 3 (Example 5.12).*

Table 5.17. *ANOVA Table (Running Data Only) for the TG Data (Example 5.12).*

Source	DF	Sum of Squares	Mean Square	F Value	Pr > F
Model	2	325.0800000	162.5400000	16.86	0.0234
Error	3	28.9200000	9.6400000		
Corrected Total	5	354.0000000			

R-Square	Coeff Var	Root MSE	Y Mean
0.918305	28.22577	3.104835	11.00000

Level of B	N	------------Y------------ Mean	Std Dev
1	2	4.1000000	1.97989899
2	2	7.7000000	2.40416306
3	2	21.2000000	4.38406204

as a one-way layout with $3 \times 3 = 9$ treatment levels, one needs only the information in Table 5.16 to construct the decision lines

$$5.08 \pm h(0.05; 9, 9)\sqrt{8.11}\sqrt{\frac{8}{18}}$$

$$\pm 3.48(1.90)$$

$$\pm 6.61$$

$$(-1.53, 11.69).$$

Table 5.18. *Cell Means for the TG Data (Example 5.12).*

Level of A	Level of B	N	-------------Y------------- Mean	Std Dev
1	1	2	−1.4500000	2.47487373
1	2	2	0.3500000	2.61629509
1	3	2	3.5500000	0.49497475
2	1	2	5.3000000	3.25269119
2	2	2	3.5000000	1.69705627
2	3	2	1.4500000	4.17193001
3	1	2	4.1000000	1.97989899
3	2	2	7.7000000	2.40416306
3	3	2	21.2000000	4.38406204

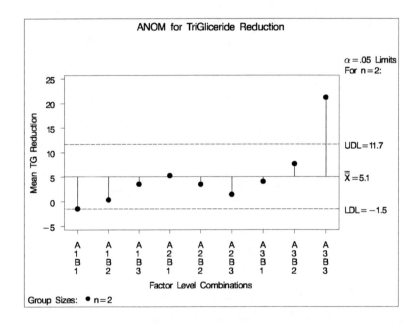

Figure 5.24. *ANOM Chart for the TG Data: One-Way Layout (Example 5.12).*

To complete the analysis, one needs the cell means, which are given in Table 5.18. The decision chart is in Figure 5.24, from which one observes that only the combination of running with the drug is significantly different from the other factor level combinations.

Methods for assessing the appropriateness of the analysis are model dependent. The analysis by means of the $J = 3$ ANOM decision charts associated with the simple effects analysis is based on model (5.11). The three sets of residuals (which are available from the three one-way ANOVAs, as, for example, from the ANOVA producing Table 5.17) should be examined for departures from normality. (See Figure 5.25 for a normal probability plot for the $A = 1$ case.) Comparison of the cell variances, as, for example, those appearing in Table 5.17, will suggest whether the HOV assumption is violated. The analysis of the

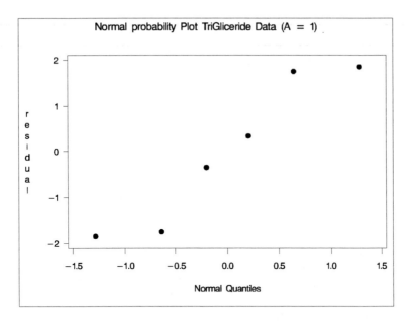

Figure 5.25. *Normal Probability Plot for the TG Data: A = 1 (Example 5.12).*

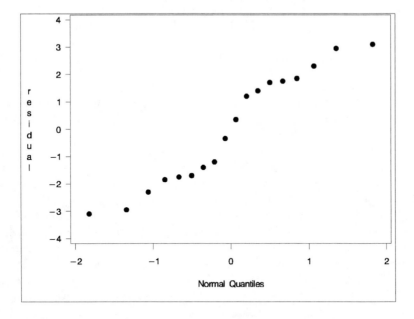

Figure 5.26. *Normal Probability Plot for the TG Data: One-Way Layout (Example 5.12).*

one-way layout was based on model (5.12), so with this choice one would examine the residuals from this model. Figure 5.26 has a normal probability plot of the residuals. (Note the residuals are the same as the residuals from fitting the ANOVA model associated with Table 5.16.)

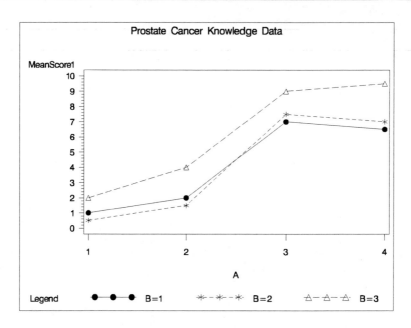

Figure 5.27. *Means Interaction Plot for PKS Data (Example 5.13).*

The HOV assumption can be assessed by carefully comparing the standard deviations in Table 5.18 or by performing a test (such as ANOMV). In short, checking the model assumptions follows exactly those steps described in Chapter 2, since in both cases, the problems break down into one or more one-way analyses.

Example 5.13 (Prostate Cancer Knowledge Data). The goal for this study was to determine factors associated with patient knowledge of prostate cancer (particularly among at-risk groups). This was a small study conducted by a fellow in a medical university clinic setting. The response is the score (higher indicates greater knowledge) on a prostate knowledge survey (PKS), which was collected in a stratified convenience sample collected at a clinic. The analysis was used to design a prostate screening grant proposal. Two factors of interest were factor A, highest level of education attained (1 = < high school, 2 = high school, 3 = some college, 4 = college graduate), and factor B, race/ethnicity (1 = black, 2 = hispanic, 3 = white). An interaction means plot (Figure 5.27) offered little evidence of interaction, which is confirmed by the p-value of 0.8475 for the significance of the A*B term in the ANOVA output in Table 5.19. Since interaction can be excluded from the analysis, two main effects of ANOM analyses will be used to summarize the results. With a statistical package with ANOM capability, these charts can be generated directly. From the chart in Figure 5.28, one can see that at $\alpha = 0.05$ the education effect is significant. (Those with education of high school or less are below the lower decision line, and those with some college or more plot above the upper decision line.) The ANOM chart for the race/ethnicity effect (Figure 5.29) reveals that this effect is significant. (Factor level 3, white, plots above the upper decision line.) Note that the ANOM analysis quantifies the visual analysis that one can make based on Figure 5.27.

Table 5.19. *ANOVA Table for the PKS Data (Example* 5.13*).*

Source	DF	Sum of Squares	Mean Square	F Value	Pr > F
Model	11	239.4583333	21.7689394	20.90	<.0001
Error	12	12.5000000	1.0416667		
Corrected Total	23	251.9583333			

R-Square	Coeff Var	Root MSE	Y Mean
0.950389	21.29991	1.020621	4.791667

Source	DF	Type I SS	Mean Square	F Value	Pr > F
A	3	215.4583333	71.8194444	68.95	<.0001
B	2	21.3333333	10.6666667	10.24	0.0025
A*B	6	2.6666667	0.4444444	0.43	0.8475

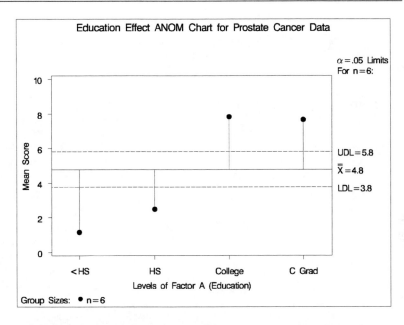

Figure 5.28. *ANOM Chart for Education Level (Example* 5.13*).*

With a statistical package without ANOM capability, the decision lines can be constructed using $MS_e = 1.042$ and df $= 12$ from the ANOVA table that was used to assess interaction (Table 5.19). Using (5.10), the education effect has ANOM decision lines

$$4.79 \pm h(0.05; 4, 12)\sqrt{1.042}\sqrt{\frac{3}{24}}$$

$$\pm 2.85(.360)$$

$$\pm 1.03$$

$$(3.76, 5.82).$$

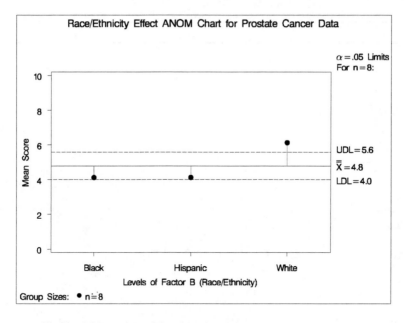

Figure 5.29. *ANOM Chart for Race/Ethnicity (Example 5.13).*

Through comparison of the factor level means for factor A (Table 5.20) to the decision limits for factor A or examination of a manually constructed chart similar to the one in Figure 5.28, one reaches the conclusions outlined previously. The race/ethnicity effect has ANOM decision lines

$$4.79 \pm h(0.05; 3, 12)\sqrt{1.042}\sqrt{\frac{2}{24}}$$

$$\pm 2.67(.295)$$

$$\pm .788$$

$$(4.00, 5.58).$$

Similarly, through comparison of the factor level means for factor B (Table 5.20) to the decision limits for factor B or examination of a manually constructed chart similar to the one in Figure 5.29, one reaches the conclusions outlined previously.

Example 5.14 (Radio Ad Data). An organization that serves the needs of the homeless is planning a nationwide campaign to raise money for homeless shelters. A preliminary study to determine the effectiveness of radio advertising was undertaken. Twelve radio markets of about the same size were selected for the study. An equivalent dollar value of advertising time was bought in each market, and total dollar contributions, Y, were tallied for each ad market. Station format and ad focus were thought to be associated with contribution level. Factor A, station format, was at four levels (1 = talk, 2 = country, 3 = rock, and 4 = religious). Factor B, ad focus, was at three levels (1 = help people, 2 = get homeless

Table 5.20. *Factor Level Means for the PKS Data (Example 5.13).*

Level of		------------Y------------	
A	N	Mean	Std Dev
1	6	1.16666667	0.98319208
2	6	2.50000000	1.76068169
3	6	7.83333333	0.98319208
4	6	7.66666667	1.50554531
Level of		------------Y------------	
B	N	Mean	Std Dev
1	8	4.12500000	2.94897076
2	8	4.12500000	3.48209707
3	8	6.12500000	3.48209707

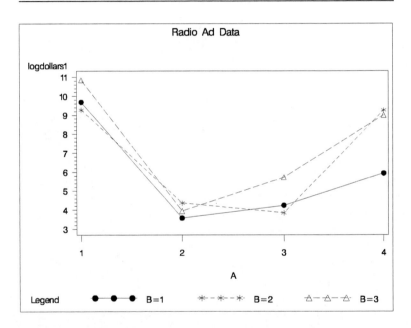

Figure 5.30. *Interaction Plot for Radio Ad Data (Example 5.14).*

off streets, and 3 = reduce crime). Factor level combinations were randomly assigned to the markets, so that one station represented each factor level combination. For example, if the combination (country/help people) was randomly assigned to Battle Creek, then a country station was used in Battle Creek with the ad focused on the help-people theme. Analysis of this design requires that there be no interaction between factors A and B. Examination of the interaction plot (Figure 5.30) suggests little evidence of interaction. (The analysis is based on the natural logarithm of dollar contributions, ln Y, since this transformation results in normally distributed residuals.) Note that one cannot test for interaction statistically and that on heuristic grounds one might have reason to suspect interaction in this application.

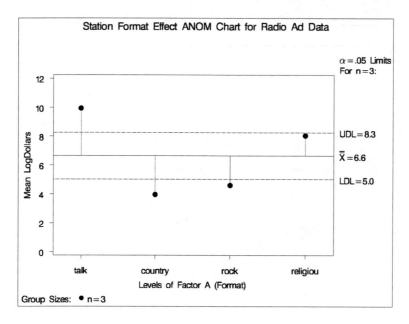

Figure 5.31. *ANOM Chart for Station Format (Example 5.14).*

Assuming no interaction, we will produce an ANOM chart for each main effect. Using a statistical package with ANOM capability, one can directly produce charts for each main effect, from which one can see that the station format (Figure 5.31) is significant at $\alpha = 0.05$ since factor level 1 (talk) plots above the upper decision limit and factor levels 2 (country) and 3 (rock) plot below the lower decision limit. Ad focus (Figure 5.32) is not significant since all three factor level means plot within the decision lines.

Using a package without ANOM capability, ANOVA with no interaction term (output in Table 5.21) can be used to calculate $MS_e = 0.963$ and identify the associated df $= 6$, thereby allowing for calculation of the decision limits

$$6.64 \pm h(0.05; 4, 6)\sqrt{0.963}\sqrt{\frac{3}{12}}$$
$$\pm 3.31(.491)$$
$$\pm 1.63$$
$$(5.01, 8.27)$$

for the station format effect and decision limits

$$6.64 \pm h(0.05; 3, 6)\sqrt{0.963}\sqrt{\frac{2}{12}}$$
$$\pm 3.07(.400)$$
$$\pm 1.23$$
$$(5.41, 7.87)$$

for the ad focus effect.

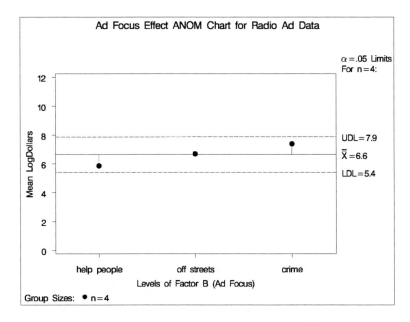

Figure 5.32. *ANOM Chart for Ad Focus (Example* 5.14).

Table 5.21. *ANOVA Table for Radio Ad Data (Example* 5.14).

Source	DF	Sum of Squares	Mean Square	F Value	Pr > F
Model	5	77.17289738	15.43457948	16.02	0.0020
Error	6	5.77925307	0.96320885		
Corrected Total	11	82.95215045			

R-Square	Coeff Var	Root MSE	logy Mean
0.930330	14.76999	0.981432	6.644773

Comparison of the factor level means in Table 5.22 to the appropriate decision limits or construction of charts similar to those in Figures 5.31 and 5.32 leads to the conclusions previously outlined.

5.4 Two-Factor ANOM for Binomial and Poisson Data

Two-factor ANOM methods for binomial and Poisson data will be illustrated with two examples. The approach is the same as in the previous section. That is, first, interaction is investigated. Based on that finding, either main effects are tested (no interaction) or the approach in step 3A (Section 5.3) can be followed.

Example 5.15 (Length of Stay Data). The length of stay (LOS) data given in Table 5.23 arises from a factorial design with $I = 2$ and $J = 3$. Hospital LOS for patients receiving

Table 5.22. *Factor Level Means for Radio Ad Data (Example 5.14).*

Level of		------------log(y)------------	
A	N	Mean	Std Dev
1	3	9.94293055	0.81446288
2	3	3.96807017	0.39323455
3	3	4.60849686	0.98878801
4	3	8.05959602	1.84236298
Level of		------------log(y)------------	
B	N	Mean	Std Dev
1	4	5.86559466	2.73713796
2	4	6.68831321	2.98498006
3	4	7.38041234	3.11690768

Table 5.23. *Data From a Factorial Design to Study Length of Stay after Bypass Surgery (Example 5.15).*

Drug	Treatment		
	1	2	3
A	10(0.20)	6(0.12)	20(0.40)
B	15(0.30)	16(0.32)	12(0.24)

bypass surgery was determined, and subjects were classified as either having LOS greater than 6 days or not. Factor 1 is the administration of a drug (two levels), and factor 2 is type of postsurgical treatment (three levels). Fifty subjects ($n = 50$) for each combination of factor levels were chosen. The counts in the cells of Table 5.23 are the number of subjects with LOS exceeding 6 days (considered the target LOS); the values in parentheses are the sample proportions. A binomial model for the counts will be adopted; that is, the cell counts are binomial with $n = 50$ and proportion with LOS greater than 6 is p_{ij} for $i = 1, 2$ and $j = 1, 2, 3$.

The first step in the analysis is the check for interaction. The plot in Figure 5.33 is an interaction plot for the LOS data and suggests significant interaction. For many researchers, this would be sufficient evidence to adopt an analysis consistent with interaction. (A statistical test for interaction will be given subsequently.) The ANOM analysis will be done by consolidating the data into a one-way layout with six levels (one for each factor level combination). The ANOM decision chart is given in Figure 5.34. Based on the decision chart, the equal proportions hypothesis cannot be rejected ($\alpha = 0.05$).

Manual construction of the ANOM decision lines requires the overall proportion of patients with LOS greater than 6 days:

$$\bar{p} = \frac{0.20 + \cdots + 0.24}{6} = 0.263.$$

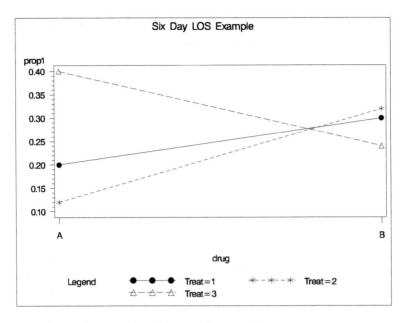

Figure 5.33. *An Interaction Plot for the Length of Stay Data (Example 5.15).*

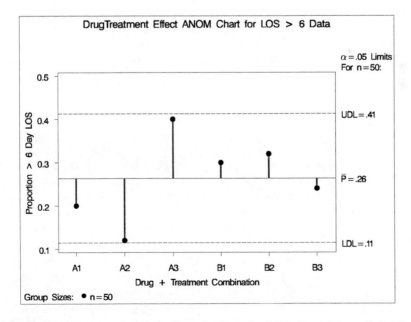

Figure 5.34. *ANOM Decision Chart for Length of Stay Data (Example 5.15).*

Table 5.24. *SAS Logistic Regression Output for Six-Day LOS Data (Example 5.15).*

	Model Fit Statistics	
		Intercept
	Intercept	and
Criterion	Only	Covariates
AIC	347.909	344.915
SC	351.613	367.138
-2 Log L	345.909	332.915

Deviance and Pearson Goodness-of-Fit Statistics				
Criterion	Value	DF	Value/DF	Pr > ChiSq
Deviance	9.4567	2	4.7284	0.0088
Pearson	9.3093	2	4.6547	0.0095

Number of unique profiles: 6

	Model Fit Statistics	
		Intercept
	Intercept	and
Criterion	Only	Covariates
AIC	347.909	350.372
SC	351.613	365.187
-2 Log L	345.909	342.372

Checking the conditions (2.7) for this example, we have

$$n\,\widehat{p}_{\min} = 50(0.12) = 6 > 5,$$
$$n(1 - \widehat{p}_{\max}) = 50(1 - 0.40) = 50(0.60) = 30 > 5.$$

Thus, the sample size is large enough, and the ANOM decision lines (2.6) with $\alpha = 0.05$ are

$$0.263 \pm h(0.05; 6, \infty)\sqrt{0.263(1 - 0.263)}\sqrt{\frac{5}{300}}$$
$$\pm 2.62(0.0569)$$
$$\pm 0.149$$
$$(0.114, 0.412).$$

One method for statistically testing for interaction with two-factor binomial data is to use logistic regression (see, e.g., Stokes et al. (2000)). In logistic regression, the logit (the log odds of the p_{ij}) is modeled as a linear function of the factors. To test for interaction, first fit the full model (one with the two main effects and interaction, which in this example requires six parameters), then fit the reduced model (one with only the two main effects, which in this case requires four parameters). For each model, the deviance (identified as $-2\mathrm{Log}L$ in the SAS output in Table 5.24) is calculated and from that the difference in the deviances is found ($342.372 - 332.915 = 9.4567$). This difference can then be tested using

Table 5.25. *Data from a Factorial Design for ER Visits by Intervention and Payor (Example 5.16).*

Intervention	Payor			
	PPO	HMO	Medicaid	All Payors
Nurse	10(0.20)	30(0.60)	14(0.28)	54(0.360)
No Nurse	30(0.60)	60(1.20)	40(0.80)	130(0.867)
Both Interventions	40(0.40)	90(0.90)	54(0.54)	184 (0.613)

a chi-squared test. In this example, the main effects model has two fewer parameters than the full model, so the df for the chi-squared test for the reduction in the deviance is 2. Since the p-value for the chi-squared test is 0.0088, one concludes that interaction is significant, which confirms the conclusion arrived at by means of the interaction plot in Figure 5.33.

Example 5.16 (Emergency Room Visits Data). The emergency room (ER) visits data in Table 5.25 arises from a factorial design with $I = 2$ and $J = 3$. The study investigated the effect of nurse intervention (home nurse assigned versus home nurse not assigned) on the ER utilization rates for patients with COPD (a serious lung disease). One hundred subjects were selected for each of three payor groups (PPO, HMO, and Medicaid); each payor group was randomly divided between those receiving home nurse visits and those not receiving home nurse visits. That is, 50 subjects ($n = 50$) for each combination of factor level were chosen. The counts in the cells of Table 5.25 are the number of ER visits for a 1-year period; the values in parentheses are the sample ER utilization rates. A Poisson model for the counts will be adopted; that is, the cell counts are Poisson with $n = 50$, and the ER utilization rate is u_{ij} for $i = 1, 2$ and $j = 1, 2, 3$.

Significant interaction is not suggested by the plot in Figure 5.35. For many researchers, this would be sufficient evidence to adopt an analysis consistent with no interaction; that is, a main effects analysis should be adopted. (A statistical test for interaction will be given subsequently.) The ANOM decision chart (Figure 5.36) for the payor main effect indicates a significant payor effect ($\alpha = 0.01$) in which PPO clients have significantly lower ER utilizations rates and HMO clients have significantly higher utilization rates. The decision chart in Figure 5.37 indicates that the intervention was successful ($\alpha = 0.01$). That is, ER utilization rates were significantly lower for COPD patients assigned a home nurse.

The main effects decision charts are constructed by combining across the other main effect and using methods described in Chapter 2. For example, the ANOM decision lines for the payor effect are based on sample sizes of $n = 100$ and sample utilization rates based on 40, 90, and 54 ER visits, respectively, for PPO, HMO, and Medicaid clients. The ANOM critical value is based on infinite df and $I = 3$ groups.

To assess interaction in a more formal manner, one can use Poisson regression. In this example, one has counts as well as a measure of exposure (number of subjects in each cell); hence, one can fit a model to the log rate with a Poisson response variable (see, e.g., Stokes et al. (2000)). (Since this is a generalized linear model, the model can be fit using PROC GENMOD in SAS.)

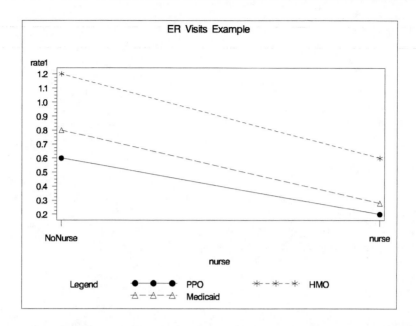

Figure 5.35. *An Interaction Plot for the ER Visits Data (Example 5.16).*

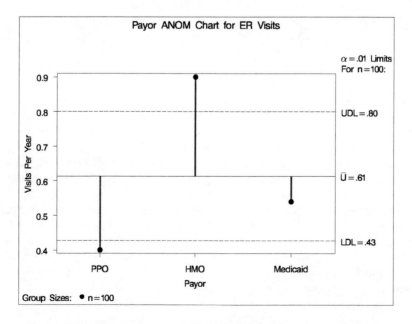

Figure 5.36. *ANOM Decision Chart for Payor Effect (Example 5.16).*

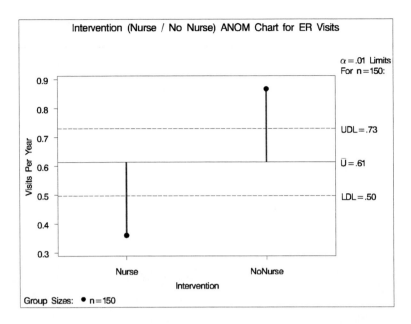

Figure 5.37. *ANOM Decision Chart for Nurse Intervention Effect (Example 5.16).*

From the SAS output in Table 5.26 one finds the goodness-of-fit statistic is the deviance (1.3614), which after division by the df (1.3614/2 = 0.6807) is approximately chi-squared distributed. This chi-squared value is clearly not statistically significant; hence the main effects model is satisfactory. Note that the parameter estimates in the generalized linear model are consistent with the ANOM decision chart (e.g., the log rate for nurse is less than that of no nurse since the parameter estimate of −0.8786 is negative).

5.5 ANOM for Higher-Order Layouts

For three-way and higher-order layouts the analysis would start (as it did with a two-way layout) by testing for interactions. Tests for three-way and higher-order interactions are best done using the ANOVA. Tests can be constructed using ANOM-type procedures, but in these cases the quantities one would end up plotting have no physical meaning, and they would be of no use in judging practical significance. If there are no significant interactions, the main effects of the factors can be studied using the ANOM decision lines (for a factor at f levels)

$$\overline{y}_{\cdot\cdots} \pm h(\alpha; f, \nu_\epsilon)\sqrt{\text{MS}_e}\sqrt{\frac{f-1}{N}}, \tag{5.18}$$

where ν_ϵ is the df for the MS_e and $\overline{y}_{\cdot\cdots}$ is the overall mean. (The exact number of dots will depend on the number of factors.)

Table 5.26. *SAS Poisson Regression Output for ER Visits (Example 5.16).*

```
                 The GENMOD Procedure
                   Model Information
       Data Set                  WORK.ERVISITS
       Distribution                    Poisson
       Link Function                       Log
       Dependent Variable              ERvisits
       Offset Variable                   ltotal


       Number of Observations Read          6
       Number of Observations Used          6
```

```
           Class Level Information
       Class      Levels   Values
       nurse         2     nurse NoNurse
       payor         3     PPO HMO Medicaid
```

Criteria For Assessing Goodness Of Fit			
Criterion	DF	Value	Value/DF
Deviance	2	1.3614	0.6807
Scaled Deviance	2	1.3614	0.6807
Pearson Chi-Square	2	1.3591	0.6795
Scaled Pearson X2	2	1.3591	0.6795
Log Likelihood		472.5796	

Algorithm converged.

				Analysis Of Parameter Estimates			
			Standard	Wald 95% Confidence		Chi-	Pr>
Parameter		DF	Estimate	Error	Limits		Square	ChiSq
Intercept		1	-0.2704	0.1441	-0.5529	0.0121	3.52	0.0606
nurse	nurse	1	-0.8786	0.1619	-1.1959	-0.5612	29.45	<.0001
nurse	NoNurse	0	0.0000	0.0000	0.0000	0.0000	.	.
payor	PPO	1	-0.3001	0.2086	-0.7090	0.1088	2.07	0.1503
payor	HMO	1	0.5108	0.1721	0.1735	0.8482	8.81	0.0030
payor	Medicaid	0	0.0000	0.0000	0.0000	0.0000	.	.
Scale		0	1.0000	0.0000	1.0000	1.0000		

NOTE: The scale parameter was held fixed.

Example 5.17 (Quality of Life Data). COPD is a debilitating lung disease. In this study, factors that affect quality of life (QOL) were assessed. The measurement instrument is a QOL survey (higher scores correspond to higher QOL). The 54 subjects enrolled in the study were stratified by gold stage disease severity (factor A at 3 levels: 18 each classified as 1 = mild IIA, 2 = mild IIB, and 3 = severe). Subjects were treated with life skills counseling (factor B at 3 levels: 1 = none, 2 = a training packet containing literature, 3 = a set of couseling sessions) and job training (factor C at 3 levels: 1 = none, 2 = a class on job search methods, 3 = specific job training). This is a three-factor study with each factor at three levels. Two subjects were assigned to each factor level combination.

Table 5.27. *ANOVA Table for the QOL Data (Example 5.17).*

Source	DF	Type I SS	Mean Square	F Value	Pr > F
A	2	5593.037037	2796.518519	22.42	<.0001
B	2	38.259259	19.129630	0.15	0.8586
A*B	4	260.518519	65.129630	0.52	0.7203
C	2	1471.814815	735.907407	5.90	0.0075
A*C	4	225.296296	56.324074	0.45	0.7704
B*C	4	97.407407	24.351852	0.20	0.9388
A*B*C	8	778.148148	97.268519	0.78	0.6241
Error	27	3368.50000	124.75926		

Table 5.28. *Means Table for the QOL Data (Example 5.17).*

R-Square	Coeff Var	Root MSE	Y Mean
0.715330	18.30521	11.16957	61.01852

Level of A	N	Mean	Std Dev
1	18	73.0555556	12.2304933
2	18	61.8333333	11.9027926
3	18	48.1666667	8.7059817

Level of B	N	Mean	Std Dev
1	18	59.9444444	15.1714601
2	18	61.1111111	13.3939797
3	18	62.0000000	16.8592792

Level of C	N	Mean	Std Dev
1	18	56.2222222	14.5543830
2	18	58.5555556	14.5044505
3	18	68.2777778	13.6847083

From the ANOVA results in Table 5.27, it is clear that there are no significant interactions. Using information from Table 5.28 as well, the ANOM decision lines (formula (5.18)) for disease severity are

$$61.02 \pm h(0.05; 3, 27)\sqrt{124.76}\sqrt{\frac{2}{54}}$$
$$\pm 2.48(2.149)$$
$$\pm 5.33$$
$$(55.69, 66.35).$$

Since factor level 1 (mild IIa) plots above the upper decision line and level 3 (severe) plots below the lower decision line, disease severity is significant (Figure 5.38). The ANOM

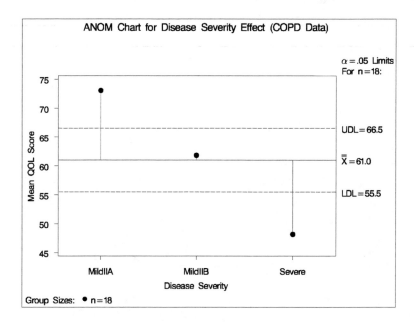

Figure 5.38. *ANOM Chart for Disease Severity (Example 5.17).*

Table 5.29. *Data from a Factorial Design to Study Hemoglobin Levels in Females (Example 5.18).*

		Drug		
		Drug 1	Drug 2	Drug 3
		12.1	13.2	15.4
		12.3	10.5	15.8
	Chemo	12.3	12.8	13.2
		8.8	12.8	13.9
Therapy		10.1	9.7	16.0
		8.7	7.7	11.1
		9.4	12.0	8.4
	Radiation	10.1	8.4	14.2
		8.6	7.6	12.2
		9.7	8.1	13.3

decision lines are the same for counseling and job training. Levels of the former are not statistically significant. Job training is significant, since the average for specific job training plots above the upper decision line.

Example 5.18 (Hemoglobin Data). Recall the hemoglobin study from Example 5.1. Suppose that the complete study included 30 females (Table 5.29) in addition to the 30 males in

Table 5.30. *ANOVA Table for the Hemoglobin Level Data (Example 5.18).*

Source	DF	Sum of Squares	Mean Square	F Value	Pr > F
Model	11	259.5738333	23.5976212	10.08	<.0001
Error	48	112.4160000	2.3420000		
Corrected Total	59	371.9898333			

R-Square	Coeff Var	Root MSE	Y Mean
0.697798	12.37321	1.530359	12.36833

Source	DF	Type I SS	Mean Square	F Value	Pr > F
A	1	96.01350000	96.01350000	41.00	<.0001
B	2	82.50533333	41.25266667	17.61	<.0001
A*B	2	4.11600000	2.05800000	0.88	0.4219
C	1	71.06816667	71.06816667	30.35	<.0001
A*C	1	0.14016667	0.14016667	0.06	0.8078
B*C	2	5.16133333	2.58066667	1.10	0.3405
A*B*C	2	0.56933333	0.28466667	0.12	0.8858

Table 5.1. The response of interest is still hemoglobin level, but the study can now be viewed as one with three factors: therapy (A), drug (B), and gender (C). From the ANOVA results in Table 5.30, it is clear that there are no significant interactions. The ANOM decision lines (5.18) for drug are

$$12.368 \pm h(0.05; 3, 48)\sqrt{2.342}\sqrt{\frac{2}{60}}$$
$$\pm 2.43(0.279)$$
$$\pm 0.679$$
$$(11.69, 13.05),$$

and the drug means are $\bar{y}_{1\bullet\bullet\bullet} = 11.475$, $\bar{y}_{2\bullet\bullet\bullet} = 11.605$, and $\bar{y}_{3\bullet\bullet\bullet} = 14.025$. From the ANOM chart in Figure 5.39, one sees that there are significant differences due to drug 3 producing significantly ($\alpha = 0.05$) higher hemoglobin levels and drugs 1 and 2 producing lower levels than the overall hemoglobin level.

The ANOM decision lines for therapy and gender are the same because they have the same number of levels. For $\alpha = 0.05$ they are

$$12.368 \pm h(0.05; 2, 48)\sqrt{2.342}\sqrt{\frac{1}{60}}$$
$$\pm 2.02(0.198)$$
$$\pm 0.400$$
$$(11.97, 12.77),$$

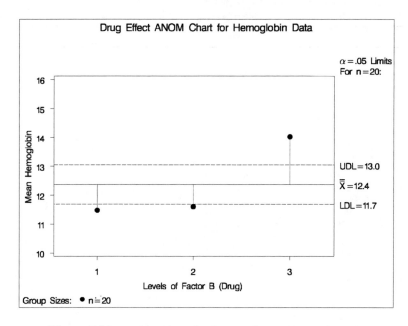

Figure 5.39. *ANOM Chart for Drug Effect in Example* 5.18.

and for $\alpha = 0.001$ they are

$$12.368 \pm h(0.001; 2, 48)\sqrt{2.342}\sqrt{\frac{1}{60}}$$
$$\pm 3.55(0.198)$$
$$\pm 0.701$$
$$(11.68, 13.06).$$

From the ANOM charts in Figures 5.40 and 5.41, one sees that therapy is significant at $\alpha = 0.001$ because chemo ($\overline{y}_{.2..} = 13.63$) is associated with significantly high hemoglobin levels and radiation ($\overline{y}_{.1..} = 11.10$) is associated with significantly low hemoglobin levels. Gender is significant at $\alpha = 0.001$ because males have significantly high hemoglobin levels ($\overline{y}_{..1.} = 13.46$) and females have significantly low hemoglobin levels ($\overline{y}_{..2.} = 11.28$).

Example 5.19 (Sales Training Data). In this study, factors affecting the effectiveness of sales training were investigated. The training is administered to large classes. The response variable is the percent of those in the training class that reach their sales quota at the end of the first 90 days subsequent to the training. Factor A, content, was at three levels corresponding to the training orientation (1 = motivational, 2 = product familiarization, and 3 = organizational). Factor B, center, was at two levels corresponding to the center's location (1 = West Coast, 2 = East Coast). Factor C, duration of training, was at two levels (1 = 1 week, 2 = 2 days) The results are given in Table 5.31. From the ANOVA results in Table 5.32, one sees that there is a significant AB interaction. Therefore, one cannot average the results for A over the different levels of B (or the results of B over the different levels of

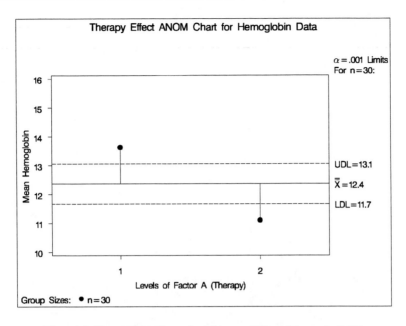

Figure 5.40. *ANOM Chart for Therapy Effect (Example 5.18).*

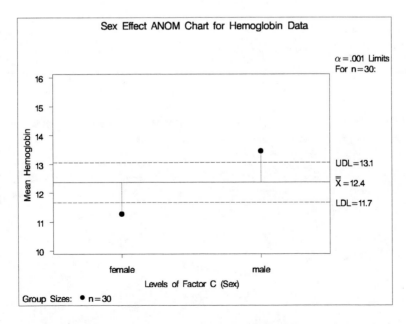

Figure 5.41. *ANOM Chart for Gender Effect (Example 5.18).*

Table 5.31. *Percent Reaching Quota for Content/Center/Duration Combinations (Example 5.19).*

		Duration			
		1 Week		2 Days	
		West		East	
		1	2	1	2
Content	1	89.9	88.3	90.6	88.2
		91.5	89.4	89.1	87.1
	2	88.4	92.3	85.7	90.4
		90.1	93.7	87.6	91.8
	3	89.5	93.1	87.5	89.7
		87.6	91.4	85.8	91.5

Table 5.32. *ANOVA for the Sales Training Data (Example 5.19).*

SOURCE	DF	SS	MS	F	P
A (A)	2	2.25083	1.12542	0.89	0.4353
B (B)	1	23.2067	23.2067	18.39	0.0011
T (C)	1	17.0017	17.0017	13.48	0.0032
A*B	2	47.8758	23.9379	18.97	0.0002
A*C	2	1.52583	0.76292	0.60	0.5621
B*C	1	0.06000	0.06000	0.05	0.8310
A*B*C	2	0.27750	0.13875	0.11	0.8968
RESIDUAL	12	15.1400	1.26167		
TOTAL	23	107.338			

A). One could study factor A separately for each level of B, but in this case it makes more sense to consider the (A, B) combinations as levels of a new factor (call it D) and study factors D and C together using the model

$$\boxed{\overline{Y}_{ijk} = \mu + \alpha_i^D + \alpha_j^C + \alpha_{ij}^{DC} + \epsilon_{ijk}}\;. \tag{5.19}$$

This results in Table 5.33, from which it is clear that there is no DC interaction. The ANOM decision lines for factor D are

$$89.59 \pm h(0.01;\, 6,\, 12)\sqrt{1.262}\sqrt{\frac{5}{24}}$$

$$\pm\, 3.98(0.513)$$

$$\pm\, 2.04$$

$$(87.55,\, 91.63),$$

Table 5.33. *ANOVA Table for the Sales Training Data Using Model* (5.19) *(Example 5.19).*

SOURCE	DF	SS	MS	F	P
(A,B) (D)	5	73.3333	14.6667	11.62	0.0003
T (C)	1	17.0017	17.0017	13.48	0.0032
D*C	5	1.86333	0.37267	0.30	0.9063
RESIDUAL	12	15.1400	1.26167		
TOTAL	23	107.338			

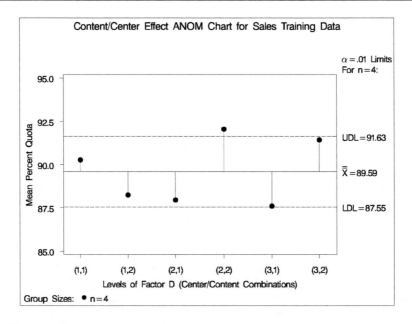

Figure 5.42. *ANOM Chart for Content/Center Combinations (Example 5.19).*

and for factor C they are

$$89.59 \pm h(0.05; 2, 12)\sqrt{1.262}\sqrt{\frac{1}{24}}$$

$$\pm 2.18(0.229)$$

$$\pm 0.500$$

$$(89.09, 90.09).$$

The ANOM charts are given in Figures 5.42 and 5.43, from which one sees that the content/center combinations are significant ($\alpha = 0.01$) due to the (2, 2) combination (product familiarization, east center) resulting in a significantly high percent reaching their quota, and duration is significant ($\alpha = 0.05$) due to 1 week of training producing a significantly high percent reaching quota and 2 days of training producing a significantly low percent reaching quota.

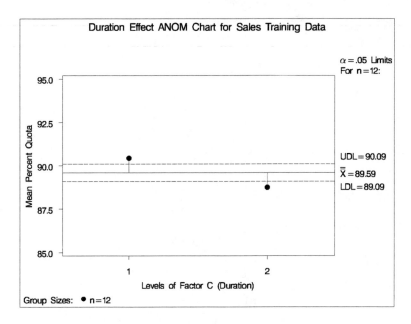

Figure 5.43. *ANOM Chart for Duration (Example 5.19).*

Table 5.34. *Paint Application Data (Example 5.20).*

	Operator 1		Operator 2		Operator 3	
	Applicator		Applicator		Applicator	
Brand	New	Old	New	Old	New	Old
1	22.5	16.9	13.9	6.0	16.7	12.3
2	23.1	17.9	17.7	8.9	19.3	9.1

Example 5.20 (Paint Application Data).   A producer of desks uses a finish (i.e., paint) on desk tops. Paint coatings are applied and then dried. An experiment is performed to find what (if any) are the effects of applicator type (new or old) and varnish brand (1 or 2) on the rework rate in the painting process. The measured response is Y = % paint jobs per hour requiring rework. Factor A is the type of applicator (at two levels), and factor B is the varnish brand (also at two levels). Three operators (factor C) will be used to run the paint process. The operators are blocks. Each operator worked for 1 hour with each combination of applicator type and particle size. The resulting data are given in Table 5.34.
    The model for these data is

$$Y_{ijk} = \mu + \alpha_i^A + \alpha_j^B + \alpha_{ij}^{AB} + \alpha_k^C + \epsilon_{ijk}\,, \tag{5.20}$$

and the ANOVA table for this model is given in Table 5.35. It is clear that the brand and the applicator type do not interact, the brand has no effect, and the applicator type has a large effect. The operator (the blocking variable) is also significant. Hence, to reduce the rework rate, the first order of business would be to use the better of the two applicator types.

Table 5.35. *ANOVA Table for the Paint Application Data Using Model* (5.20) *(Example* 5.20).

Source	DF	Sum of Squares	Mean Square	F Value	Pr > F
Model	5	305.7008333	61.1401667	21.80	.0009
Error	6	16.8283333	2.8047222		
Corrected Total	11	22.5291667			

R-Square	Coeff Var	Root MSE	Y Mean
0.947824	10.90438	1.674730	15.35833

Source	DF	Type I SS	Mean Square	F Value	Pr > F
A	1	147.7008333	147.7008333	52.66	0.0003
B	1	4.9408333	4.9408333	1.76	0.2327
C	2	149.7516667	74.8758333	26.70	0.0010
A*B	1	3.3075000	3.3075000	1.18	0.3192

Of secondary importance would be to look at how the operators are running the applicators differently from one another. Decision lines for applicator are

$$15.36 \pm h(0.001; 2, 6)\sqrt{2.805}\sqrt{\frac{1}{12}}$$
$$\pm 5.96(0.483)$$
$$\pm 2.879$$
$$(12.48, 18.24),$$

and for operator they are

$$15.36 \pm h(0.05; 3, 6)\sqrt{2.805}\sqrt{\frac{2}{12}}$$
$$\pm 3.07(0.684)$$
$$\pm 2.099$$
$$(13.26, 17.45).$$

From the ANOM charts in Figures 5.44 and 5.45, one sees that applicator 1 results in a significantly high ($\alpha = 0.001$) rework rate, while applicator 2 results in a significantly low rework rate. Operator 1 produces significantly more ($\alpha = 0.05$) rework paint jobs than average, and operator 2 produces significantly fewer rework paint jobs than average. If one considers the blocks (operators) as random effects, then a somewhat different analysis may be conducted using mixed models. The operator ANOM decision chart is useful only for comparing these particular operators.

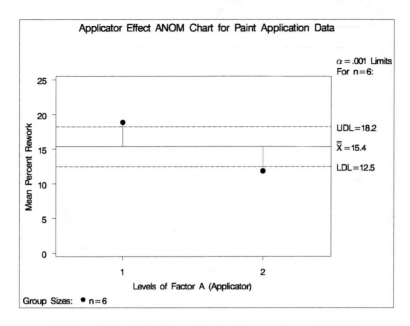

Figure 5.44. *ANOM Chart for Applicator (Example 5.20).*

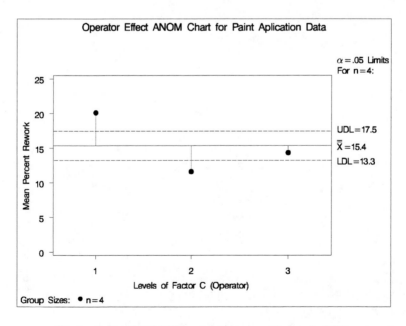

Figure 5.45. *ANOM Chart for Operator (Example 5.20).*

Chapter 5 Problems

1. Consider an experiment with three factors (A, B, C), each at two levels (high, low), and with six observations per cell. Cell means and variances are given in the table below.

	Means A(L)		Means A(H)		Variances A(L)		Variances A(H)	
	C(L)	C(H)	C(L)	C(H)	C(L)	C(H)	C(L)	C(H)
B(H)	16.36	11.43	20.57	15.03	5.40	3.91	8.14	2.72
B(L)	12.09	7.76	18.09	14.33	2.17	1.58	3.20	4.31

(a) First, analyze the experimental data for only the low level of A (i.e., as a 2 × 2 experiment) using an interaction plot, testing for interaction, and, if appropriate, testing for main effects.

(b) Next, analyze the experimental data for only the high level of A.

(c) Finally, analyze the complete experiment.

(d) Repeat the analysis, assuming there were only two observations per cell.

2. A certain brand and model of lawn mower can be fitted with either an expensive or a cheap carburetor. A store manager wants to know what ratio of oil-gasoline (high or low) should be used with each carburetor to maximize fuel efficiency. Sixteen lawn mowers were selected at random from stock. Eight of these were randomly fitted with the cheap carburetor, and the remaining eight were fitted with the expensive one. From each group of eight mowers, four were selected at random to be used with the high ratio, and the remaining four were used with the low ratio. One gallon of the proper mixture was put in each machine, and the length of time in minutes the lawn mower ran before exhausting the fuel was recorded. Results are in the table below. What did the experiment reveal? Propose an alternative design that still uses 16 runs but uses only four lawn mowers.

Carburetor	Oil Ratio Low				Oil Ratio High			
Expensive	42	43	40	45	36	39	32	39
Cheap	15	25	27	22	38	32	34	34

3. A biochemist wishes to compare four compounds used to remove nicotine from chewing tobacco. The following experiment is performed. Four chaws of tobacco are each cut into four pieces (producing a total of 16 pieces), and then the smaller pieces are separated, and each compound is randomly applied to one piece within each chaw by immersing the separated piece for 2 minutes in a beaker containing the compound. A standard method is then used to determine the remaining nicotine level in each piece. Data are shown below. What important design principle was employed here? What is it about this problem that makes it likely to be useful? Reanalyze the data as a complete layout.

	chaw			
compound	1	2	3	4
comp1	34.11	35.19	37.70	39.86
comp2	36.39	40.12	44.14	43.37
comp3	34.20	36.30	40.71	39.84
comp4	38.18	41.12	43.13	46.01

4. A researcher interested in the direct marketing of drugs to senior citizens considers three factors that might affect responses to the advertisements: message clarity, type size, and the level of risk communicated in the ad. Each factor has two levels (low, high). Eight ads are constructed (one at each factor level combination). Two responses are of interest: comprehension (measured by the subject's score on a 20-question true-false test) and the subject's indication (on a seven-point Likert scale where 7 corresponds to highly likely) of the likelihood the subject would ask a doctor to prescribe the drug. Forty persons from an upscale retirement home in Florida volunteer to participate in the study. Five volunteers are randomly assigned to each of eight tables which contain the appropriate ad copy. After exposure to the ad, each subject fills out a questionnaire and takes the 20-question test. The data are in the table below (the likelihood of asking for the prescription is in parentheses). Analyze the data. Do you see any problems with the Likert scale data?

	Clarity(L)		Clarity(H)	
	Type(L)	Type(H)	Type(L)	Type(H)
risk(H)	16(5)	18(7)	11(7)	9(2)
	13(3)	16(7)	13(6)	12(1)
	17(4)	15(7)	10(5)	8(3)
	17(2)	16(4)	11(6)	8(3)
	16(4)	18(7)	10(4)	14(5)
risk(L)	8(5)	9(7)	12(5)	18(3)
	9(2)	16(7)	13(3)	17(2)
	8(3)	11(7)	12(2)	16(2)
	11(3)	12(7)	9(2)	19(1)
	10(4)	11(4)	11(5)	20(1)

5. Suppose that in problem 2 there are only four lawn mowers of the brand and type of interest. Suppose that each lawn mower is equipped with the expensive carburetor. Each of the four lawn mowers is tested with each combination of carburetor and fuel level. The data are shown below. It was randomly determined which carburetor should be used on each run. Is that a good idea? Analyze the data. Compare this design with the design in problem 2.

carburetor	oil	mower 1	mower 2	mower 3	mower 4
cheap	low	38.63	30.28	36.71	46.34
expensive	high	42.89	38.37	40.49	47.63
cheap	low	39.60	33.79	39.95	45.21
expensive	high	45.91	42.25	45.57	45.98

6. Five subjects agree to participate in a study to compare four formulations of a topical pain killer. The subject is blindfolded, the pain killer is applied to a 1 inch^2 area of skin, and after 60 seconds a pinprick test is used to determine pain threshold. A run number is assigned to each subject formulation combination, and the runs are done in random order. The person conducting the test is unaware of which formulation is being used on a run. The reading (see table below) is the pin depth required to elicit a pain response. Are the formulations different in their effectiveness?

formulation	subject 1	subject 2	subject 3	subject 4	subject 5
1	48	91	74	69	95
2	36	51	32	42	32
3	76	127	86	94	43
4	23	50	38	59	81

7. Failure of transient HIV-positive patients to show up for treatment in AIDS clinics is a problem. Eighteen clinics participated in the following study. Two factors were used to determine outcome. One factor was a reward: no reward, money, fast food coupons. The other factor was information content at previous appointment: no information, risk to patient stressed, risk to public stressed. The data below list the percentage of no-shows for appointments in a 30-day period at two clinics for each treatment level combination. Analyze the data.

	no information	patient	public
no reward	24.1,31.2	19.1,24.3	39.8,24.7
money	14.2,22.5	28.8,25.4	24.1,32.1
coupons	31.5,23.1	17.8,18.0	23.0,29.5

Chapter 6

Incomplete Multifactor Studies

While the complete layouts discussed in the previous chapter allow one to estimate all possible interactions, a heavy price is paid in terms of the number of observations required. For example, even with only three levels for each of three factors and $n = 2$ replicates, $(3)(3)(3)(2) = 54$ experimental runs are required. In this chapter we consider four specific kinds of incomplete designs (which require fewer runs)—Latin squares, Graeco–Latin squares, balanced incomplete block designs, and Youden squares. Each is potentially useful in a different experimental situation.

6.1 Latin Squares

Consider the problem of comparing different brands of tires for tread wear. Since both the particular car being driven and the position of the tire would potentially affect the tread wear, one would want to consider both car and position as blocking factors. For this purpose, a design such as the one given in Table 6.1 would be appropriate. Note that each brand appears once and only once in each position (row) and once and only once on each car (column). Such a design is called a Latin square and is a special incomplete three-way layout.

What is gained by using a Latin square design is that one can study three factors, each at M levels with only M^2 rather than M^3 observations. There are, however, two restrictions. First, the factors must all have the same number of levels; second, one must assume that there is no interaction between the factors. Specifically, one assumes the model

$$Y_{ijk} = \mu + \alpha_i^A + \alpha_j^B + \alpha_k^C + \epsilon_{ijk} \quad , \tag{6.1}$$

where the ϵ_{ijk} are independent $N(0, \sigma^2)$ and the (i, j, k) combinations belong to a set $\mathcal{D}$ (denoted $(i, j, k) \in \mathcal{D}$), which depends on the particular Latin square. The particular Latin square should be chosen at random. This is accomplished by first choosing a standard square at random (such as in Table 6.1, where the first row and first column are in order) and then randomizing the columns and all but the first rows. Tables of standard squares can be found in Fisher and Yates (1963) and Wu and Hamada (2000).

Table 6.1. *A* 4×4 *Latin Square for Comparing Four Brands* (a, b, c, d) *of Tires with Regard to Tread Wear.*

		Car			
		1	2	3	4
	1	a	b	c	d
Position	2	b	c	d	a
	3	c	d	a	b
	4	d	a	b	c

Table 6.2. *Tread Wear (0.1 mm) for Four Brands* (a, b, c, d) *of Tires Using a Latin Square Design Together with Column Means, Row Means, and the Grand Mean (Example 6.1).*

		Car								
		1		2		3		4		
	1	a	8	c	11	d	2	b	8	7.25
Tire	2	c	7	a	5	b	2	d	4	4.5
Position	3	d	3	b	9	a	7	c	9	7
	4	b	4	d	5	c	9	a	3	5.25
		5.5		7.5		5		6		6

For model (6.1) the residuals are

$$
\begin{aligned}
\widehat{\epsilon}_{ijk} &= y_{ijk} - \widehat{\mu} - \widehat{\alpha}_i^A - \widehat{\alpha}_j^B - \widehat{\alpha}_k^C \\
&= y_{ijk} - \overline{y}_{\cdots} - (\overline{y}_{i\cdot\cdot} - \overline{y}_{\cdots}) - (\overline{y}_{\cdot j\cdot} - \overline{y}_{\cdots}) - (\overline{y}_{\cdot\cdot k} - \overline{y}_{\cdots}) \\
&= y_{ijk} - \overline{y}_{i\cdot\cdot} - \overline{y}_{\cdot j\cdot} - \overline{y}_{\cdot\cdot k} + 2\overline{y}_{\cdots},
\end{aligned}
\tag{6.2}
$$

and the MS_e is

$$
MS_e = \frac{\sum_{(i,j,k)\in\mathcal{D}} \left(\widehat{\epsilon}_{ijk}\right)^2}{(M-1)(M-2)}
\tag{6.3}
$$

with $(M-1)(M-2)$ df. The ANOM decision lines (5.18) can be used to test for main effects. (See Nelson (1993) for the mathematical details.)

Example 6.1 (Tread Wear Data). The results of a tire tread wear experiment designed using a Latin square with cars and tire position as blocks are shown in Table 6.2. The design (the letters), row means, column means, and grand mean are also included in the table.

The residuals computed using (6.2) are given in Table 6.3. Note that calculating the residuals requires the brand means (Table 6.4). Using (6.3), one obtains $MS_e = 4.167$, and a normal probability plot of the residuals (Figure 6.1) suggests that the assumptions of normality and equal variances are reasonable. One could use a general linear model routine in a software package (fitting a main effects model with no interaction) to

Table 6.3. *Residuals for the Tread Wear Data (Example 6.1).*

			1		2		3		4
					Car				
	1	a	1.5	c	−0.75	d	−1.75	b	1
Tire	2	c	0	a	−0.75	b	−1.25	d	2
Position	3	d	−1	b	0.75	a	1.25	c	−1
	4	b	−0.5	d	0.75	c	1.75	a	−2

Table 6.4. *Tread Wear (0.1 mm) for Four Brands* (a, b, c, d) *of Tires Rearranged for Easy Computation of Brand Means (Example 6.1).*

			Car			
		1	2	3	4	
	a	8	5	7	3	5.75
Brand	b	4	9	2	8	5.75
	c	7	11	9	9	9
	d	3	5	2	4	3.5
		5.5	7.5	5	6	6

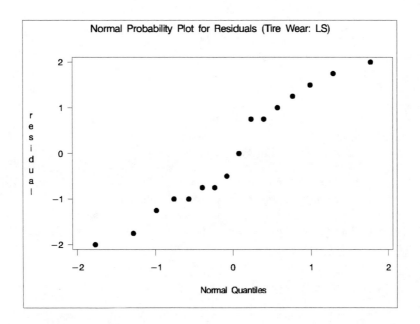

Figure 6.1. *Normal Probability Plot of Residuals for the Tread Wear Data (Example 6.1).*

Table 6.5. *General Linear Model Output for the Tread Wear Data (Example 6.1).*

```
Dependent Variable: wear
```

| | | Sum of | | | |
Source	DF	Squares	Mean Square	F Value	Pr > F
Model	9	97.000	10.7777778	2.59	0.1299
Error	6	25.000	4.1666667		
Corrected Total	15	122.000			

R-Square	Coeff Var	Root MSE	wear Mean
0.795082	34.02069	2.041241	6.000000

Least Squares Means	
brand	wear LSMEAN
a	5.75000000
b	5.75000000
c	9.00000000
d	3.50000000

obtain MS_e, the df, and the group means (see Table 6.5) for easy insertion into the decision lines formula. The residuals from this fitted model can be used to check for normality.

The ANOM decision lines for brand are

$$6 \pm h(0.05; 4, 6)\sqrt{4.167}\sqrt{\frac{3}{16}}$$

$$\pm 3.31(0.884)$$

$$\pm 2.93$$

$$(3.07, 8.93),$$

and from the ANOM chart (Figure 6.2), one sees that there are significant ($\alpha = 0.05$) differences due to brand c being significantly bad.

In the tread wear example, the rows and columns were both blocking factors, but any three factors (with the same number of levels) could be studied using a Latin square design if the assumption of no interactions were reasonable. Unfortunately, in most physical science and engineering experiments, that is not the case, and Latin square designs should be used with great caution since the presence of interactions can invalidate the results.

6.2   Graeco–Latin Squares

Two $M \times M$ Latin squares (one denoted with Roman letters and one denoted with Greek letters) are said to be orthogonal if, when they are superimposed, every one of the M^2 letter combinations appears only once. An example is shown in Table 6.6. Such a design is a

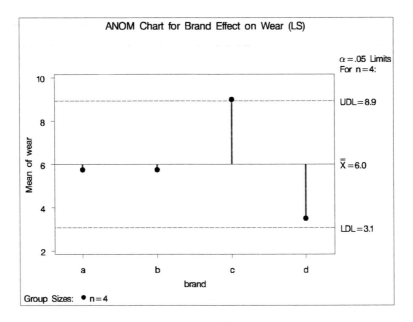

Figure 6.2. *ANOM Chart for the Tread Wear Data (Example 6.1).*

Table 6.6. *A 4×4 Graeco–Latin Square.*

aα	cγ	dδ	bβ
cδ	aβ	bα	dγ
dβ	bδ	aγ	cα
bγ	dα	cβ	aδ

special incomplete four-way layout and is called a Graeco–Latin square. It can be used to study four factors, each at M levels with only M^2 observations, if, as with the Latin square, none of the factors interact. Specifically, one assumes the model

$$Y_{ijkl} = \mu + \alpha_i^A + \alpha_j^B + \alpha_k^C + \alpha_l^D + \epsilon_{ijkl} \quad , \qquad (6.4)$$

where the ϵ_{ijkl} are independent $N(0, \sigma^2)$ and $(i, j, k, l) \in \mathcal{D}$. Obviously, all the cautions regarding Latin squares also apply to Graeco–Latin squares. Fisher and Yates (1963) provided sets of $M - 1$ orthogonal Latin squares of size M (a set of Latin squares is called orthogonal if every pair in the set is orthogonal) for $M = 3, 4, 5, 7, 8, 9$ (no 6×6 orthogonal Latin squares exist). One should choose a Graeco–Latin square by choosing two orthogonal Latin squares at random, superimposing them, and then randomizing the columns and all but the first rows. For model (6.4) the residuals are

$$\begin{aligned}
\widehat{\epsilon}_{ijkl} &= y_{ijkl} - \widehat{\mu} - \widehat{\alpha}_i^A - \widehat{\alpha}_j^B - \widehat{\alpha}_k^C - \widehat{\alpha}_l^D \\
&= y_{ijkl} - \overline{y}_{....} - (\overline{y}_{i...} - \overline{y}_{....}) - (\overline{y}_{.j..} - \overline{y}_{....}) \\
&\quad - (\overline{y}_{..k.} - \overline{y}_{....}) - (\overline{y}_{...l} - \overline{y}_{....}) \\
&= y_{ijkl} - \overline{y}_{i...} - \overline{y}_{.j..} - \overline{y}_{..k.} - \overline{y}_{...l} + 3\overline{y}_{....},
\end{aligned} \tag{6.5}$$

and the MS_e is

$$MS_e = \frac{\sum_{(i,j,k,l)\in\mathcal{D}}\left(\widehat{\epsilon}_{ijkl}\right)^2}{(M-1)(M-3)} \tag{6.6}$$

with $(M-1)(M-3)$ df. The ANOM decision lines (5.18) can be used to test for main effects (see Nelson (1993)).

Example 6.2 (Tread Wear Data). Suppose that in the tire tread wear experiment considered in Example 6.1, four mechanics participated and the Graeco–Latin square design given above was used. (The Greek letters represent the four mechanics.) To minimize the amount of additional computation, assume that the results of the experiment are the same as those given in Example 6.1. Using (6.5) and (6.6), one can compute $MS_e = 7.167$ with 3 df, and the ANOM decision lines (5.18) for brand (or any of the other factors) with $\alpha = 0.1$ are

$$6 \pm h(0.1; 4, 3)\sqrt{7.167}\sqrt{\frac{3}{16}}$$
$$\pm 3.50(1.159)$$
$$\pm 4.06$$
$$(1.94, 10.06).$$

From the ANOM chart in Figure 6.3 one sees that there is no effect due to brand. As in the Latin square example, one can use a generalized linear model routine to obtain the necessary MS_e for the ANOM decision lines. The output from such a routine is given in Table 6.7.

6.3　Balanced Incomplete Block Designs

In a randomized block design, the size of the block must equal the number of treatments (the number of levels of the factor of interest), as in Example 5.11, in which the blocks were of size three and three formulations were being compared. Sometimes it is convenient (or even necessary) to use block sizes that are smaller than the number of treatments, as, for example, if one were interested in comparing five brands of tires and using cars (four tires per car) as blocks. For that situation, balanced incomplete block (BIB) designs are useful. A BIB design is a design in which

1. each treatment is replicated the same number of times,

2. all blocks are the same size,

3. no treatment appears twice in the same block, and

4. each pair of treatments appears in the same number of blocks. (This is the "balanced" part of the design.)

Let

$$I = \text{number of treatments,}$$
$$J = \text{number of blocks,}$$
$$n = \text{number of replicates,}$$
$$b = \text{block size.}$$

A BIB design with $I = 5$, $J = 5$, $n = 4$, and $b = 4$ is shown in Table 6.8. Tables of BIB designs can be found in Fisher and Yates (1963), Davies (1978), and Box, Hunter, and Hunter (1978). Once a design has been chosen, everything possible should be randomized. Namely, numbers should be assigned to the blocks at random, letters should be assigned to the treatments at random, and the treatments within the blocks should be randomized.

With a BIB design, one assumes the model

$$Y_{ij} = \mu + \alpha_i^A + \alpha_j^B + \epsilon_{ij} \tag{6.7}$$

for the $N = In = Jb$ pairs $(i, j) \in \mathcal{D}$ and that the ϵ_{ij} are independent $N(0, \sigma^2)$. Factor A represents the treatments, and factor B represents the blocks. This is actually a two-way

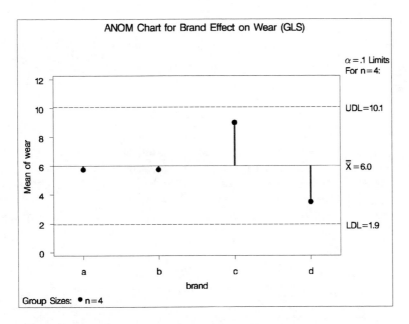

Figure 6.3. *ANOM Chart for the Tread Wear Data Using a Graeco–Latin Square (Example* 6.2*).*

Table 6.7. *General Linear Model Output for the Tread Wear Data (Example 6.2).*

```
Dependent Variable: wear
```

Source	DF	Sum of Squares	Mean Square	F Value	Pr > F
Model	12	100.500	8.3750000	1.17	0.5098
Error	3	21.500	7.1666667		
Corrected Total	15	122.000			

R-Square	Coeff Var	Root MSE	wear Mean
0.823770	44.61772	2.677063	6.000000

Least Squares Means	
brand	wear LSMEAN
a	5.75000000
b	5.75000000
c	9.00000000
d	3.50000000

Table 6.8. *A BIB Design for Five Treatments* (a, b, c, d, e) *in Blocks of Size Four.*

		Block		
1	2	3	4	5
a	a	b	a	a
b	b	c	c	b
c	d	d	d	c
d	e	e	e	e

design with many missing values for which, because of its structure, it is possible to compute directly estimates of the α_i^A, which are needed for an exact test of H_A. These estimates must be adjusted (often referred to as "adjusted for blocks" or "eliminating blocks") because of the missing values, and hence are computed differently than in a balanced complete design (i.e., they are not of the form $\bar{y}_{i\cdot\cdot} - \bar{y}_{\cdot\cdot\cdot}$). For BIB designs, the estimators of the treatment effects α_i^A are (see Nelson (1993))

$$\widehat{\alpha}_i^A = \frac{I-1}{J(b-1)} \mathcal{G}_i \,, \tag{6.8}$$

where

$$\mathcal{G}_i = g_i - T_i/b = \text{the } i\text{th adjusted treatment total} \tag{6.9}$$
$$= g_i - (T - T_i')/b, \tag{6.10}$$

$g_i = i$th treatment total,

$T = $ sum of all the observations,

$T_i = $ sum of the block totals in which the ith treatment appears,

$T_i' = $ sum of the block totals in which the ith treatment is absent

$\quad = T - T_i.$

The $\mathcal{G}_i$ are most conveniently computed using (6.9) if $2n > J$ and using (6.10) otherwise. The estimators for the block effects are computed in a similar fashion. Namely,

$$\widehat{\alpha}_j^B = \frac{J-1}{I(n-1)} \mathcal{H}_i \,, \tag{6.11}$$

where

$$\mathcal{H}_j = h_j - S_j/n = \text{the } j\text{th adjusted block total} \tag{6.12}$$

$$\quad = h_j - (T - S_j')/n \tag{6.13}$$

$h_j = j$th block total,

$S_j = $ sum of the treatment totals for treatments in the jth block,

$S_j' = $ sum of the treatment totals for treatments not in the jth block,

$\quad = T - S_j.$

The $\mathcal{H}_j$ are most conveniently computed using (6.12) if $2b > I$ and using (6.13) otherwise. Residuals for model (6.7) are

$$\widehat{\epsilon}_{ij} = y_{ij} - \overline{y}_{..} - \widehat{\alpha}_i^A - \widehat{\alpha}_j^B \,, \tag{6.14}$$

and the MS_e is

$$\mathrm{MS}_e = \frac{\sum_{(i,j)\in\mathcal{D}} \left(\widehat{\epsilon}_{ij}\right)^2}{N - I - J + 1} \tag{6.15}$$

with df $= N - I - J + 1$.

Because the $\widehat{\alpha}_i^A$ are not of the usual form, the usual ANOM procedure is not applicable. The $\widehat{\alpha}_i^A$ are still normally distributed, are still unbiased estimators, and still have correlations $\rho = -1/(I-1)$. However, their variances are (see Nelson (1993))

$$\mathrm{Var}(\widehat{\alpha}_i^A) = \sigma^2 \frac{(I-1)^2}{IJ(b-1)}.$$

It follows that under hypothesis H_A, the joint distribution of the

$$T_i = \frac{\widehat{\alpha}_i^A}{\sqrt{\mathrm{MS}_e}} \frac{\sqrt{IJ(b-1)}}{I-1}$$

is the same multivariate t as was used to obtained the ANOM critical values $h(\alpha; I, \nu)$. Therefore, the same critical values are appropriate for BIB designs, and the ANOM decision lines for the $\widehat{\alpha}_i^A$ are

$$\boxed{\; 0 \pm h(\alpha; I, \nu_\epsilon)\sqrt{\mathrm{MS}_e}\, \frac{I-1}{\sqrt{IJ(b-1)}} \;} \,. \tag{6.16}$$

Table 6.9. *Average Access Times (ms) for Five Brands of Half-Height Fixed Disk Drives Together with Row Totals, Column Totals, and the Grand Total (Example 6.3).*

		1	2	3	4	5	
	a	35	41	×	32	40	148
	b	42	45	40	×	38	165
Brand	c	31	×	42	33	35	141
	d	30	32	33	35	×	130
	e	×	40	39	36	37	152
		138	158	154	136	150	736

(Blocks across the top: 1, 2, 3, 4, 5)

Example 6.3 (Disk Drive Data). Five different brands of half-height fixed disk drives were to be compared for their average access time. To eliminate any differences due to the computer in which they were used, it was desired to install samples of all the brands on the same machine. Unfortunately, the machines available would accommodate no more than four half-height drives. Therefore, it was decided to use the BIB design in Table 6.8. The resulting average access times in milliseconds (ms) are given in Table 6.9.

Using (6.10), one computes

$$\mathcal{G}_a = 148 - (736 - 154)/4 = 2.5,$$

$$\mathcal{G}_b = 165 - (736 - 136)/4 = 15,$$

$$\mathcal{G}_c = 141 - (736 - 158)/4 = -3.5,$$

$$\mathcal{G}_d = 130 - (736 - 150)/4 = -16.5,$$

$$\mathcal{G}_e = 152 - (736 - 138)/4 = 2.5.$$

In this example, we have $I = 5$, $J = 5$, and $b = 4$; therefore, one obtains (equation (6.8))

$$\widehat{\alpha}_1^A = 0.67,$$

$$\widehat{\alpha}_2^A = 4,$$

$$\widehat{\alpha}_3^A = -0.93,$$

$$\widehat{\alpha}_4^A = -4.4,$$

$$\widehat{\alpha}_5^A = 0.67.$$

Similarly, for the block effects one obtains (equation (6.13))

$$\mathcal{H}_a = 138 - (736 - 152)/4 = -8,$$

$$\mathcal{H}_b = 158 - (736 - 141)/4 = 9.25,$$

$$\mathcal{H}_c = 154 - (736 - 148)/4 = 7,$$

$$\mathcal{H}_d = 136 - (736 - 165)/4 = -6.75,$$

$$\mathcal{H}_e = 150 - (736 - 130)/4 = -1.5,$$

Table 6.10. *Residuals for the Disk Drive Data Using the BIB Design in Table* 6.8.

		\multicolumn{5}{c}{Blocks}				
		1	2	3	4	5
	a	−0.3334	1.0666	×	−3.6667	2.9333
	b	3.3333	1.7333	−2.6667	×	−2.4000
Brand	c	−2.7334	×	4.2666	−1.0667	−0.4667
	d	−0.2667	−2.8667	−1.2667	4.4000	×
	e	×	0.0666	−0.3334	0.3333	−0.0667

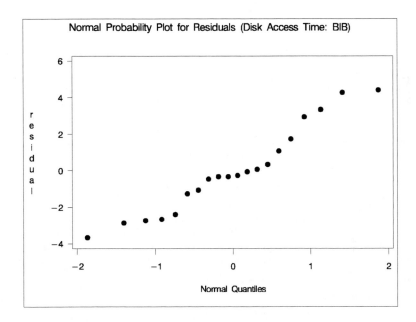

Figure 6.4. *Normal Probability of the Residuals for the Disk Drive Data (Example* 6.3*).*

and (equation (6.11))

$$\widehat{\alpha}_1^B = -2.133,$$
$$\widehat{\alpha}_2^B = 2.466,$$
$$\widehat{\alpha}_3^B = 1.866,$$
$$\widehat{\alpha}_4^B = -1.8,$$
$$\widehat{\alpha}_5^B = -0.4.$$

Using (6.3), one can compute the residuals in Table 6.10. The normal probability plot of the residuals given in Figure 6.4 indicates no difficulty with the assumption of normal

Table 6.11. *General Linear Model Output for the Disk Drive Data (Example 6.3).*

```
Dependent Variable: time
```

| | | Sum of | | | |
Source	DF	Squares	Mean Square	F Value	Pr > F
Model	8	234.4000000	29.3000000	3.02	0.0466
Error	11	106.8000000	9.7090909		
Corrected Total	19	341.2000000			

R-Square	Coeff Var	Root MSE	time Mean
0.686987	8.467232	3.115941	36.80000

Least Squares Means	
brand	time LSMEAN
a	37.4666667
b	40.8000000
c	35.8666667
d	32.4000000
e	37.4666667

errors, and from Table 6.10 (using (6.15)) one obtains

$$\text{MS}_e = \frac{106.8}{20 - 5 - 5 + 1} = 9.71$$

with df $= 11$. Alternatively, one can obtain necessary quantities for the ANOM decision lines, including MS_e, from the main effects model as well as the brand effect estimates using the adjusted (least squares) means from Table 6.11 and the grand mean. For example, $\widehat{\alpha}_1^A = 37.467 - 36.80 = 0.67$.

The $\alpha = 0.05$ ANOM decision lines (6.16) are

$$0 \pm h(0.05; 5, 11)\sqrt{9.71}\frac{4}{\sqrt{(5)(5)(3)}}$$

$$\pm 3.01(1.439)$$

$$\pm 4.33$$

$$(-4.33, 4.33).$$

Comparing the $\widehat{\alpha}_i^A$ with these decision lines, one finds (see Figure 6.5) that brand d is significantly ($\alpha = 0.05$) fast.

6.4   Youden Squares

Suppose that in Example 6.3 one were concerned that the positions of the disk drives in the machine would have an effect on their performance, perhaps because of temperature variation within the machine. In a situation like this, one would like a design where each

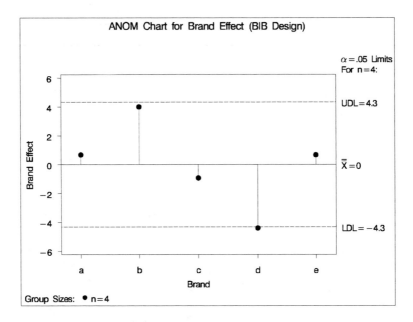

Figure 6.5. *ANOM Chart for the Disk Drive Data (Example 6.3).*

Table 6.12. *A Youden Square with $J = 5$ Blocks and $b = 4$ Positions within the Blocks Obtained from a 5×5 Latin Square.*

		Block				
		1	2	3	4	5
Position	1	a	d	b	c	e
	2	b	e	c	d	a
	3	c	a	d	e	b
	4	d	b	e	a	c
		e	c	a	b	d	⟵ Discard

drive appeared once in each position. Designs having the same properties as BIB designs, and where in addition each treatment appears once in each position within the block, are called Youden squares. A Youden square is not actually a square but an incomplete Latin square, where position within the block is the third factor. If $b(b-1)/(J-1)$ is an integer, then a Youden square with J blocks can be obtained from a $J \times J$ Latin square by discarding $J - b$ rows. Table 6.12 is an example of this with $J = 5$ and $b = 4$. Analysis for a Youden square design is the same as that for a BIB design except that one has position (within the block) as a third factor. The model for this situation is

$$Y_{ijk} = \mu + \alpha_i^A + \alpha_j^B + \alpha_k^C + \epsilon_{ijk} \quad , \tag{6.17}$$

Table 6.13. *Residuals for the Disk Drive Data Using the Youden Square in Table* 6.12.

		1	2	3	4	5
			Blocks			
	a	1.0666	2.0666	×	−3.0667	−0.6667
	b	0.3333	2.3333	−1.2667	×	−1.4000
Brand	c	−1.7334	×	1.2666	0.3333	0.1333
	d	0.3333	−1.4667	−0.2667	1.4000	×
	e	×	−2.9334	0.2666	1.3333	1.3333

and the corresponding residuals are

$$\widehat{\epsilon}_{ijk} = y_{ijk} - \overline{y}_{\bullet\bullet\bullet} - \widehat{\alpha}_i^A - \widehat{\alpha}_j^B - \widehat{\alpha}_k^C, \tag{6.18}$$

where the $\widehat{\alpha}_i^A$ and $\widehat{\alpha}_j^B$ are computed using (6.8) and (6.11), respectively, and

$$\widehat{\alpha}_k^C = \overline{y}_{\bullet\bullet k} - \overline{y}_{\bullet\bullet\bullet}.$$

The MS_e is

$$\mathrm{MS}_e = \frac{\sum_{(i,j,k)\in\mathcal{D}} \left(\widehat{\epsilon}_{ijk}\right)^2}{N - I - J - b + 2} \tag{6.19}$$

with $N - I - J - b + 2$ df. The ANOM decision lines (6.16) are also appropriate for this design (see Nelson (1993)).

Example 6.4 (Disk Drive Data). Suppose that the data on average access times for fixed disk drives given in Example 6.3 had actually been collected using the Youden square design given above. The estimates of the position effects are

$$\widehat{\alpha}_1^C = -1.4, \quad \widehat{\alpha}_2^C = 3, \quad \widehat{\alpha}_3^C = -1, \quad \widehat{\alpha}_4^C = -0.6.$$

The residuals (equation (6.18)) for model (6.17) are easily computed using the residuals from model (6.7), for example,

$$\widehat{\epsilon}_{111} = -0.3334 - (-1.4) = 1.0666.$$

All the residuals are given in Table 6.13, from which one can compute (equation (6.19))

$$\mathrm{MS}_e = \frac{45.2}{20 - 5 - 5 - 4 + 2} = 5.65$$

with 8 df. The normal probability plot of the residuals (Figure 6.6) again indicates there is no problem with the assumption of normality. The $\alpha = 0.05$ ANOM decision lines (6.16) are

$$0 \pm h(0.05; 5, 8)\sqrt{5.65}\,\frac{4}{\sqrt{(5)(5)(3)}}$$

$$\pm 3.52$$

$$(-3.52, 3.52),$$

and the ANOM chart in Figure 6.7 indicates that at $\alpha = 0.05$, brand b is significantly slow and brand d is significantly fast.

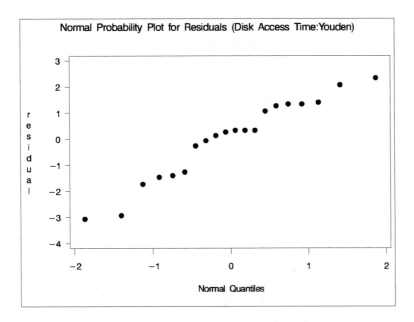

Figure 6.6. *Normal Probability Plot of the Residuals for the Disk Drive Data Using the Youden Square in Table 6.12 (Example 6.4).*

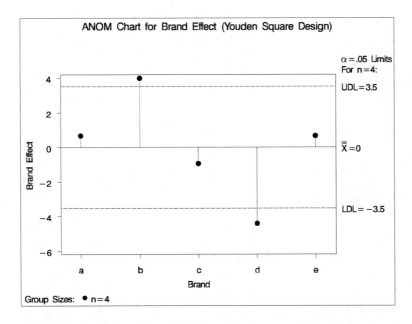

Figure 6.7. *ANOM Chart for the Disk Drive Data Using the Youden Square in Table 6.12 (Example 6.4).*

Chapter 7

Axial Mixture Designs

When the response of an experiment depends only on the proportion of each factor and not on the total amount of all the factors, then one is said to be experimenting with a mixture. Some examples of mixtures are blends of gasoline, ingredients in a pharmaceutical drug, and the components in an adhesive. Studying response surfaces for mixture experiments requires special kinds of designs or transformations, of either the factors or the standard response surface designs.

Let $x_i \geq 0$ denote the proportion of component i, where proportions are often measured by either weight, volume, or mole fraction. If there are k components (factors) in the mixture, then

$$\sum_{i=1}^{k} x_i = 1. \tag{7.1}$$

Thus, there are two major differences when it comes to designing mixture experiments. First, the levels of the different factors are not all independent; second, the experimental region is constrained. The k factors can be transformed to $k - 1$ factors whose levels are all independent, but the experimental region will (usually) still be constrained. Using the transformed factors, a standard response surface design can be chosen within the constrained experimental region, and then either the design points can be transformed back to the original factor space or the experiment and its analysis can be carried out using the transformed factors. Transformation is useful for more complicated mixture experiments in which the number of components cannot be reduced to at most three or four. For mixture experiments with fewer components, however, the special mixture designs are useful. More information on mixture experiments can be found in Cornell (2002, 1983).

When there are no constraints on the proportions x_i other than that they sum to one (equation (7.1)), then the constrained experimental region is a simplex. For two components (i.e., in two dimensions) a simplex is a line, for three components a simplex is a triangle, and for four components a simplex is a tetrahedron. With three components the proportions x_i can be conveniently plotted on triangular graph paper, as shown in Figure 7.1. The vertices of the triangle represent the three degenerate mixtures consisting of a single component, the sides of the triangle represent mixtures of only two components, and the interior points

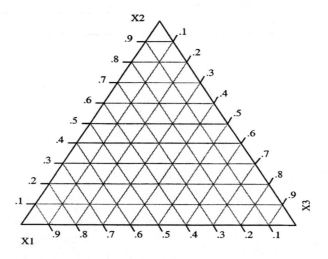

Figure 7.1. *Triangular Graph Paper.*

of the triangle represent mixtures in which all three components are present. Three useful designs for modeling over the entire simplex are simplex-lattice designs, simplex-centroid designs, and axial designs. Axial designs, however, are the only one of the three for which an ANOM procedure exists.

Axial designs are useful when one has a large number of components and wants to screen them to find the most important ones. They are designs with points positioned symmetrically on the component axes. The axis of component i is the line connecting the point $x_i = 0$ and $x_j = \frac{1}{k-1}$ for $j \neq i$ with the point $x_i = 1$ and $x_j = 0$ for $j \neq i$. Along this line, the amount of component i varies from 0 to 1, and the remaining components are maintained in equal proportions. We will consider only the simplest form of axial design in which the points are positioned equidistant (a distance Δ) from the centroid $\left(\frac{1}{k}, \ldots, \frac{1}{k}\right)$ and are located between the centroid and the vertices. A three-component axial design of this type is shown in Figure 7.2.

The estimator of the effect of component i is

$$E_i = b_i^* - \sum_{j \neq i} b_j^*/(k-1), \tag{7.2}$$

which is the difference between the estimator of the response at $x_i = 1$ and the estimator of the response at $x_i = 0$ and $x_j = \frac{1}{k-1}$ for $j \neq i$ when using the first-order model

$$Y = \beta_0 + \sum_{i=1}^{k} \beta_i x_i + \epsilon$$

$$= \beta_0 \sum_{i=1}^{k} x_i + \sum_{i=1}^{k} \beta_i x_i + \epsilon$$

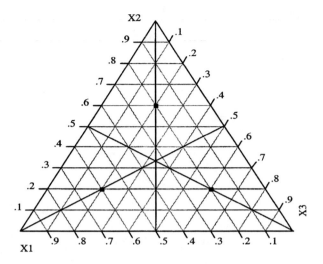

Figure 7.2. *A Three-Component Axial Design.*

$$= \sum_{i=1}^{k} (\beta_0 + \beta_i) x_i + \epsilon$$

$$= \sum_{i=1}^{k} \beta_i^* x_i + \epsilon.$$

Thus, E_i estimates the change in response when x_i changes from one to zero along the axis of component i. The relative magnitudes of the E_i indicate the relative effects of the components.

If n replicate observations are taken at each of the points in the axial design, then the statistical significance of each component effect can be tested as follows. Let

$$s_i^2 = \text{the sample variance of the responses at distance } \Delta \text{ on axis } i,$$

$$\text{MS}_e = \frac{1}{k} \sum_{i=1}^{k} s_i^2 . \tag{7.3}$$

The E_i are normal random variables, are independent of the MS_e, have correlations $\rho = -1/(k-1)$, and have variance (see Nelson (1993)),

$$\text{Var}(E_i) = \frac{k-1}{n\Delta^2 k} \sigma^2.$$

Therefore, under the null hypothesis that there are no component effects (i.e., all E_i are zero),

$$T_i = \frac{E_i}{\sqrt{\text{MS}_e}} \sqrt{\frac{n\Delta^2 k}{k-1}} \sim t(k(n-1))$$

Table 7.1. *The Coded Amounts of Tomatoes Produced Using Fertilizers Consisting of Various Mixtures of Five Ingredients and an Axial Design (Example 7.1).*

Axis	Plant 1	Plant 2	$\bar{y}_i$	s_i^2
1	7.0	9.8	8.4	3.92
2	23.0	20.2	21.6	3.92
3	11.6	9.0	10.3	3.38
4	24.1	27.3	25.7	5.12
5	14.0	17.0	15.5	4.5

and $(T_1, \ldots, T_k)$ has the same multivariate t distribution as was used to obtain the critical points $h(\alpha; k, \nu)$. Thus, the ANOM critical points $h(\alpha; k, \nu)$ can be used with axial mixture designs, and the ANOM decision lines for the component effects (similar to the decision lines (6.16)) are

$$0 \pm h(\alpha; k, k(n-1))\sqrt{\text{MS}_e}\sqrt{\frac{k-1}{n\Delta^2 k}}. \tag{7.4}$$

An effect E_i is statistically significant (at level α) if it falls outside the decision lines (7.4).

To compute the E_i, one must compute the estimates b_i^*, which can be easily obtained by numerically solving the system of equations

$$\bar{y}_i = \Delta^* b_i^* + \frac{1-\Delta^*}{k-1}\sum_{j \neq i} b_j^*, \quad i = 1, \ldots, k, \tag{7.5}$$

where

$$\bar{y}_i = \text{the average response at distance } \Delta \text{ on axis } i,$$

$$\Delta^* = \Delta + \frac{1}{k}.$$

Example 7.1 (Fertilizer Data). A fertilizer for tomatoes was to be developed using five possible ingredients. To determine the effects of the ingredients, a mixture experiment was run using an axial design with $\Delta = \frac{7}{15}$. Each mixture was used on two plants. The response recorded was pounds of tomatoes produced, and the coded results are given in Table 7.1. For $k = 5$ and $\Delta = \frac{7}{15}$ the system of (7.5) is

$$\bar{y}_i = \frac{2}{3}b_i^* + \sum_{j \neq i} b_j^*/12, \quad i = 1, \ldots, 5,$$

which when solved numerically yields the results in Table 7.2. Also given in Table 7.2 are

Table 7.2. *Estimates of the Model Parameters and the Component Effects (Example 7.1).*

i	b_i^*	E_i
1	2.757	-16.93
2	25.39	11.36
3	6.014	-12.86
4	32.41	20.14
5	14.93	-1.713

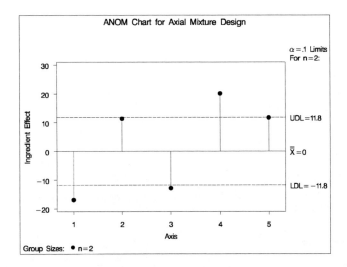

Figure 7.3. *ANOM Chart for Fertilizer (Example 7.1).*

the component effects obtained using (7.2). The decision lines (7.4) with $\alpha = 0.1$ are

$$0 \pm h(0.1; 5, 5)\sqrt{MS_e}\sqrt{\frac{4}{10/9}}$$
$$\pm (3.05)\sqrt{4.168}\sqrt{3.6}$$
$$\pm 11.8$$
$$(-11.8, 11.8),$$

where $MS_e = 4.168$ was obtained by averaging the sample variances (equation (7.3)). The ANOM chart in Figure 7.3 compares the component effects with these decision lines, and one finds that ingredients 2 and 5 do not have significant ($\alpha = 0.1$) effects.

Chapter 8

Heteroscedastic Data

In this chapter we discuss the generalization of the ANOM procedure to the heteroscedastic situation in which the process variances are not necessarily equal. In addition, this heteroscedastic analysis of means (HANOM) procedure allows an experimenter to set a goal of detecting differences among the I treatment means when two of the treatment means differ by a specified amount δ that does not depend on the process variances. See Dudewicz and Nelson (2003) for the mathematical details.

8.1 The One-Way Layout

As before, let I be the number of treatments being compared, and let y_{ij} be the jth observation from the ith population, where

$$Y_{ij} = \mu_i + \epsilon_{ij}$$

and all the observations are independent. Now, however, we assume that $\epsilon_{ij} \sim N(0, \sigma_i^2)$. Collecting the data for and performing a HANOM consists of the following steps:

1. Specify an initial sample size n_0 (≥ 2), take samples of size n_0 from each of the I populations, and calculate the sample means $\overline{y}_{0i}$ and the sample variances s_i^2.

2. Specify the level of significance α, a power γ, and the amount δ such that any two treatment means differing by δ will lead to rejection of the null hypothesis that all the treatment means are equal with power γ. From Figures A.17–A.24 (for $\alpha = 0.05$; for other α levels see Dudewicz and Nelson (2003)), find the figure with the appropriate α and I combination, and for the specified γ and df $= n_0 - 1$ find the corresponding value of w. Alternatively, one could choose a value of w using (8.1) based on a desired sample size n_i.

3. Compute
$$n_i = \max\{n_0 + 1, \lfloor (w/\delta)^2 s_i^2 \rfloor + 1\} \tag{8.1}$$

for each i, where $\lfloor y \rfloor$ denotes the greatest integer in y, and take $n_i - n_0$ additional observations $y_{i,n_0+1}, \ldots, y_{i,n_i}$ from population i.

4. For each i calculate the sample mean
$$\bar{y}_i = \frac{y_{i,n_0+1} + \cdots + y_{i,n_i}}{n_i - n_0} \tag{8.2}$$

of the second set of observations from population i.

5. For each i compute
$$b_i = \frac{n_i - n_0}{n_i}\left[1 + \sqrt{\left(\frac{n_0}{n_i - n_0}\right)\left(\left[\frac{\delta}{w}\right]^2 \frac{n_i}{s_i^2} - 1\right)}\,\right] \tag{8.3}$$

and
$$\tilde{\bar{y}}_i = (1 - b_i)\bar{y}_{0i} + b_i\bar{y}_i, \tag{8.4}$$

and then compute
$$\tilde{\bar{y}}_{\bullet} = (\tilde{\bar{y}}_1 + \cdots + \tilde{\bar{y}}_I)/I. \tag{8.5}$$

6. Compute decision lines
$$\boxed{\tilde{\bar{y}}_{\bullet} \pm \mathcal{H}(\alpha; I, n_0 - 1)\frac{\delta}{w}}\,, \tag{8.6}$$

where $\mathcal{H}(\alpha; I, n_0 - 1)$ is found in Table B.7, and reject the hypothesis $H_0 : \mu_1 = \cdots = \mu_I$ if any of the $\tilde{\bar{y}}_i$ fall outside these decision lines.

Example 8.1 (Adapted From Bishop and Dudewicz (1978)). An experiment was conducted to test the effects of different solvents on the ability of fungicide methyl-2-benzimidazole-carbamate to destroy the fungus *Penicillium expansum*. The fungicide was diluted in four different solvents and sprayed on the fungus, and the percentage of fungus destroyed was measured. The experimenter was interested in testing the hypothesis

$$H_0 : \mu_1 = \mu_2 = \mu_3 = \mu_4$$

that the four solvents resulted in the same average percentages of fungus being destroyed. Initial samples of size $n_0 = 10$ were taken with each of the four solvents. The data and summary statistics are given in Table 8.1. Using ANOMV to test for equal variances, one obtains the ANOMV chart in Figure 8.1, from which one would conclude that at the 0.05 level, the sample variance for solvent 3 is significantly large and the sample variance for solvent 4 is significantly small. Since the assumption of equal variances is not reasonable in

Table 8.1. *Data and Summary Statistics for the First-Stage Samples (Example 8.1).*

	Solvent 1		Solvent 2		Solvent 3		Solvent 4	
	95.39	98.07	92.15	94.25	95.15	94.28	95.99	96.68
	96.07	95.32	92.16	91.13	90.75	91.14	97.12	95.67
	96.58	97.41	94.75	92.09	91.97	91.45	97.07	96.49
	95.67	95.81	94.69	96.07	93.39	92.35	96.23	97.52
	96.73	97.79	93.53	96.15	86.97	94.92	97.19	95.30
$\bar{y}_{0i}$	96.48		93.70		92.24		96.53	
s_i^2	0.999		3.122		5.894		0.524	

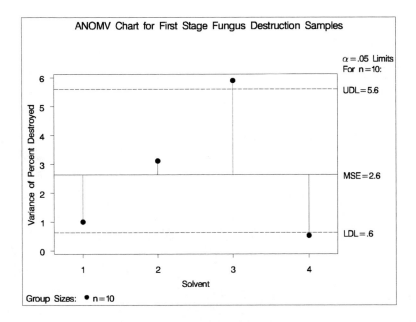

Figure 8.1. *ANOMV Chart for Example 8.1.*

this case, the HANOM procedure would be applicable. The experimenter wanted to conduct the test at level of significance $\alpha = 0.05$ and to have a power of at least $\gamma = 0.85$ if any two means differed by at least $\delta = 2.5$. From Figure A.18 ($\alpha = 0.05$ and $I = 4$), one finds that for $\gamma = 0.85$ and df $= 9$ the corresponding value of w is approximately 6. Using the sample variances from the first stage of the experiment, together with $\delta = 2.5$ and $w = 6$, one can compute the necessary values of n_i (equation (8.1)). For example,

$$n_1 = \max\{11, \lfloor (6/2.5)^2 0.999 \rfloor + 1\}$$
$$= \max\{11, 6\}$$
$$= 11.$$

Table 8.2. *Second-Stage Data and Computed Statistics (Example 8.1).*

	Solvent 1	Solvent 2		Solvent 3			Solvent 4
	94.98	96.61	95.68	94.21	92.01	94.43	96.50
	95.12	96.95	92.25	90.86	93.22	95.09	96.85
	96.37	94.39	94.98	93.59	95.01	93.57	97.54
	99.03	94.17	97.71	93.62	94.46	95.05	95.56
	95.16	95.58	93.44	97.01	96.27	99.39	98.04
	97.21	92.34	96.99	91.12	94.77	90.73	97.73
	94.45	92.75	93.01	94.72	95.53	94.68	98.00
	95.68	93.64		90.82	94.07	92.25	97.35
	96.67	98.64		91.56	95.21	92.24	97.14
	94.27	94.07		93.47	95.39	92.25	96.77
	95.78	94.23		93.35	91.68	95.55	97.30
					93.05		
$\overline{y}_i$	95.884	94.857		93.830			97.162
b_i	0.3654	0.4766		0.7235			0.5584
$\widetilde{y}_i$	96.264	94.250		93.390			96.881

Similarly, $n_2 = 18$, $n_3 = 34$, and $n_4 = 11$. Note that the stage-two sample size is proportional to the stage-one sample variance. Table 8.2 gives the second-stage data and the summary values calculated using (8.2)–(8.4). Using the values of $\widetilde{y}_i$ from Table 8.2 and (8.5), one obtains $\widetilde{y}_{\bullet} = 95.196$, and the decision lines (8.6) are

$$95.196 \pm \mathcal{H}(0.05; 4, 9)\frac{\delta}{w}$$

$$\pm 2.55\frac{2.5}{6}$$

$$\pm 1.0625$$

$$(94.134, 96.259).$$

The HANOM chart in Figure 8.2 shows that all the solvents are significantly different from the grand mean at level $\alpha = 0.05$. (Note that this chart lists $n = 10$, which was the initial sample size, not the overall sample size.) Solvents 1 and 4 destroyed significantly more fungus, and solvents 2 and 3 destroyed significantly less fungus.

8.2   Higher-Order Layouts

In designs with more than one factor, one must first test for possible interactions among the factors. As in the homogeneous variance case, this can be done using ANOVA. We will consider here only a two-way layout. Details on heteroscedastic ANOVA for a two-way layout can be found in Bishop and Dudewicz (1978), and still higher-order layouts are discussed in Bishop and Dudewicz (1981).

The usual model for the two-way layout is

$$Y_{ijk} = \mu + \alpha_i^A + \alpha_j^B + \alpha_{ij}^{AB} + \epsilon_{ijk},$$

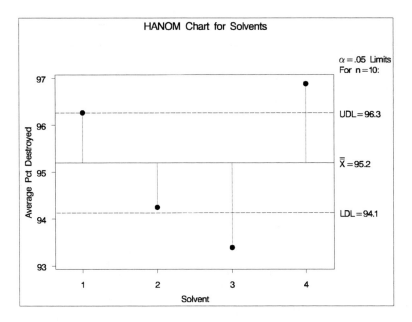

Figure 8.2. *HANOM Chart for Solvents (Example 8.1).*

where there are I levels of factor A and J levels of factor B. In the heteroscedastic case, one assumes that $\epsilon_{ijk} \sim N(0, \sigma_{ij}^2)$. Initially one would take a sample of size $n_0 \geq 2$ from each of the IJ treatment combinations and compute the usual unbiased estimates s_{ij}^2 of the σ_{ij}^2 and the averages of the observations for each treatment combination, $\overline{y}_{0ij}$. If it is known a priori that the σ_{ij}^2 are not all equal, or if a test on the variances such as the ANOMV indicates that the variances are not all equal, then heteroscedastic ANOVA and HANOM are appropriate. Sample sizes at the second stage are computed using (8.1), where s_i is replaced with s_{ij} and n_i is replaced with n_{ij}. That is,

$$n_{ij} = \max\{n_0 + 1, \lfloor (w/\delta)^2 s_{ij}^2 \rfloor + 1\}. \tag{8.7}$$

An appropriate value for w depends on whether there is significant interaction and, if there is, how one will proceed with the analysis. For fixed α, γ, and δ the value of w increases as the number of means being compared increases. Thus, a conservative choice for w would be based on comparing IJ means, which is the situation that would occur if one found significant interaction and were then interested in comparing the IJ treatment combinations. Alternatively, if one were interested in the main effects of a factor, then it would make sense to choose w based on $\max(I, J)$.

The heteroscedastic ANOVA test of $H_{AB} : \alpha_{ij}^{AB} = 0$ for all (i, j) is conducted using essentially the usual interaction sum of squares with $\overline{y}_{ij}$ replaced by $\tilde{\overline{y}}_{ij}$. Specifically, for each (i, j) one would compute

$$b_{ij} = \frac{n_{ij} - n_0}{n_{ij}} \left[1 + \sqrt{\left(\frac{n_0}{n_{ij} - n_0} \right) \left(\left[\frac{\delta}{w} \right]^2 \frac{n_{ij}}{s_{ij}^2} - 1 \right)} \right] \tag{8.8}$$

and

$$\bar{\tilde{y}}_{ij} = (1 - b_{ij})\bar{y}_{0ij} + b_{ij}\bar{y}_{ij}, \tag{8.9}$$

$$\bar{\tilde{y}}_{i\bullet} = \frac{1}{J}\sum_{j=1}^{J}\bar{\tilde{y}}_{ij}, \tag{8.10}$$

$$\bar{\tilde{y}}_{\bullet j} = \frac{1}{I}\sum_{i=1}^{I}\bar{\tilde{y}}_{ij}, \tag{8.11}$$

$$\bar{\tilde{y}}_{\bullet\bullet} = \frac{1}{IJ}\sum_{i=1}^{I}\sum_{j=1}^{J}\bar{\tilde{y}}_{ij}. \tag{8.12}$$

The hypothesis H_{AB} is tested using

$$\tilde{F}_{AB} = \left(\frac{w}{\delta}\right)^2\sum_{i=1}^{I}\sum_{j=1}^{J}(\bar{\tilde{y}}_{ij} - \bar{\tilde{y}}_{i\bullet} - \bar{\tilde{y}}_{\bullet j} + \bar{\tilde{y}}_{\bullet\bullet})^2, \tag{8.13}$$

which is compared with the appropriate quantile of an $((n_0 - 1)/(n_0 - 3))\chi^2_{(I-1)(J-1)}$ distribution.

If no significant interaction is found, then the main effects of factor A can be tested by comparing the $\bar{\tilde{y}}_{i\bullet}$ with the decision lines

$$\boxed{\bar{\tilde{y}}_{\bullet\bullet} \pm \mathcal{H}(\alpha; I, n_0 - 1)\frac{\delta}{w}}. \tag{8.14}$$

Similarly, the main effects of factor B can be tested by comparing the $\bar{\tilde{y}}_{\bullet j}$ with the decision lines

$$\boxed{\bar{\tilde{y}}_{\bullet\bullet} \pm \mathcal{H}(\alpha; J, n_0 - 1)\frac{\delta}{w}}. \tag{8.15}$$

If significant interaction is found, then one can either compare the IJ treatment combinations using the $\bar{\tilde{y}}_{ij}$ and decision lines

$$\boxed{\bar{\tilde{y}}_{\bullet\bullet} \pm \mathcal{H}(\alpha; IJ, n_0 - 1)\frac{\delta}{w}} \tag{8.16}$$

or (assuming $I > J$) compare the I levels of factor A separately for each level j' of factor B using the $\bar{\tilde{y}}_{ij'}$ and decision lines

$$\boxed{\bar{\tilde{y}}_{\bullet j'} \pm \mathcal{H}(\alpha; I, n_0 - 1)\frac{\delta}{w}}. \tag{8.17}$$

Table 8.3. *Summary Statistics from the First-Stage and Second-Stage Samples (Example 8.2).*

i	j	n_0	$\overline{y}_{0ij}$	s_{ij}^2	n_{ij}	$\overline{y}_{ij}$	b_{ij}	$\widetilde{\overline{y}}_{ij}$
1	1	6	53.467	1.8827	7	49.8	0.6184	51.199
1	2	6	55.083	2.2496	7	52.3	0.5543	53.540
1	3	6	55.250	5.827	8	57.1	0.3479	55.894
1	4	6	58.683	2.3496	7	59.0	0.5389	58.854
2	1	6	52.250	1.259	7	51.7	0.7744	51.824
2	2	6	54.250	2.187	7	54.1	0.5643	54.165
2	3	6	56.850	5.619	8	52.35	0.3799	55.140
2	4	6	59.367	3.2546	7	60.9	0.4243	60.017
3	1	6	51.333	2.6906	7	53.2	0.4914	52.250
3	2	6	53.217	3.1576	7	57.5	0.4351	55.080
3	3	6	57.433	14.4386	19	55.338	0.7245	55.915
3	4	6	59.067	1.8067	7	58.4	0.6336	58.644

Example 8.2 (Insulation Data). An experiment was conducted to study the effects of three different types of insulation (factor A) at four different temperatures (factor B) on the insulation's ability to maintain a fixed temperature. Initial samples of size six were taken for each of the 12 treatment combinations. The data are a coded version of the rise in temperature after a fixed amount of time. The sample means and variances are given in Table 8.3. The ANOMV chart in Figure 8.3 indicates that at the 0.05 level, the σ_{ij}^2 are not all equal. Thus, heteroscedastic ANOVA and HANOM are appropriate ways to continue the analysis.

The experimenters were interested in conducting tests using $\alpha = 0.1$ and being able to detect any effect of $\delta \geq 7$ with a power of 0.8. For $\alpha = 0.1$, $I = 12$, and df $= 5$, the value of w (from Dudewicz and Nelson (2002)) is approximately 8. Using (8.7) the sample sizes for the second stage were computed and are given in Table 8.3. The statistics computed from the second-stage data using (8.8)–(8.12) are also given in Table 8.3. Using (8.13), one then would compute $\widetilde{F}_{AB} = (8/7)^2(2.4855) = 3.246$, which when compared with $(5/3)\chi^2(0.05; 6) = (5/3)(12.592) = 21.0$ is found not to be significant. Since there is no significant interaction, the main effect of the two factors can be studied using HANOM. For convenience, the row means, column means, and grand mean (of the $\widetilde{\overline{y}}_{ij}$) are given in Table 8.4. For factor A one would compute decision lines (8.14)

$$55.210 \pm \mathcal{H}(0.1; 3, 5)(7/8)$$
$$\pm (2.16)(7/8)$$
$$\pm 1.89$$
$$(53.320, 57.100),$$

and from the HANOM chart in Figure 8.4, one sees that there is no effect due to the

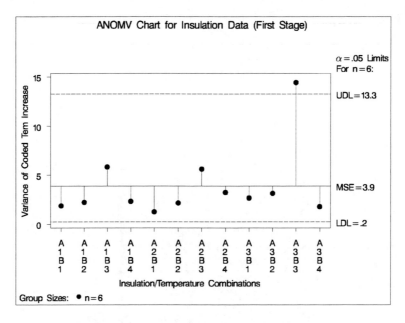

Figure 8.3. *The ANOMV Chart for Example* 8.2.

Table 8.4. *Average Outputs for Insulation-Temperature Combinations Together with Row Means, Column Means, and Grand Mean (Example 8.2).*

		Insulation			
		1	2	3	
	1	51.199	51.824	52.250	51.758
Temperature	2	53.540	54.165	55.080	54.262
	3	55.894	55.140	55.915	55.650
	4	58.854	60.017	58.644	59.172
		54.872	55.287	55.473	55.210

insulations at the 0.1 level. Similarly, for factor B one would compute decision lines (8.14)

$$55.210 \pm \mathcal{H}(0.1; 4, 5)(7/8)$$
$$\pm (2.53)(7/8)$$
$$\pm 2.21$$
$$(53.0, 57.4),$$

and from the HANOM chart in Figure 8.5, one sees that the temperatures have an effect at the 0.1 level.

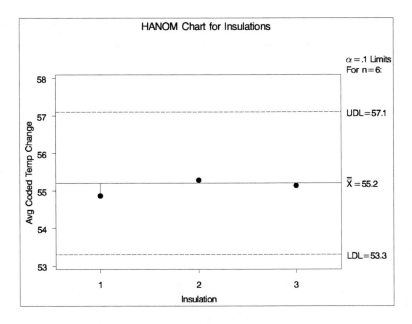

Figure 8.4. *The HANOM Chart for Insulation (Example 8.2).*

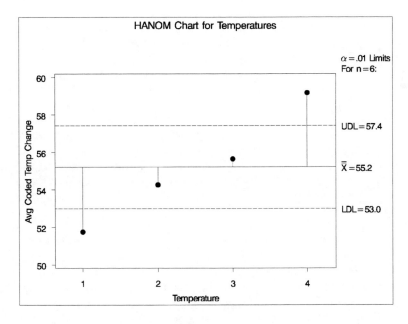

Figure 8.5. *The HANOM Chart for Temperature (Example 8.2).*

Chapter 9
Distribution-Free Techniques

A number of different ANOM-type procedures (for both means and variances) that deal with nonnormality have been investigated (see Wludyka and Nelson (1997b, 1999) and Bernard and Wludyka (2001)). These include subsampling, jackknifing, ranking, transformed ranks, and various randomization tests. Here we will discuss the methods that were found to perform the best, beginning with HOV tests.

9.1 Robust Variance Tests

Often a test for homogeneity of variances among several populations is desired. When normality cannot be safely assumed, a robust test is needed, since normal based tests such as ANOMV have unacceptably high Type I error rates for fat-tailed or skewed distributions (see, e.g., Wludyka and Nelson (1997)).

The Odd Sample Size Case

Consider the situation in which one has I independent samples of size n_i. For samples in which n_i is odd, there will be at least one ADM equal to zero for that sample. One can argue that since variability is being measured by the ADM, the median imparts no information about variability, and hence the zero ADMs should be deleted from the analysis. When this is done, the df will be reduced for each zero ADM deleted. Compensating for this, MS_e will, on average, be smaller (since removing the zero ADMs decreases the range of the ADMs). In the example that follows, the zero ADMs are deleted.

Example 9.1 (Deviation from Target Thickness Data). Four types of nozzles (A, B, C, and D) can be used to spray a polymer coating on the surface of a rectangular flange. The average thickness can be controlled by the belt speed (spray time) of the conveyor that passes through the spray chamber. Variations in the thickness of the coating are important. Four nozzles are tested by spraying 15 flanges and measuring the thickness of the coating at the center of the flange. The medians of the four nozzles are -0.005, -0.365, 0.098, and -0.194 for A, B, C, and D, respectively. The deviations from the target thickness are

163

Table 9.1. *Deviation from Target Thickness and Absolute Deviations from the Median (ADM) with Ranks in Parentheses for Four Samples of Size 15 (Example 9.1).*

Deviation from Target Thickness				ADM (Rank)			
A	B	C	D	A	B	C	D
5.217	-0.907	0.001	-0.194	5.221 (58)	0.542 (25)	0.097 (6)	0.000(2.5)
-5.779	-0.365	1.562	-1.377	5.774 (59)	0.000(2.5)	1.464 (42)	1.184 (36)
0.161	0.620	-1.924	-0.629	0.166 (10)	0.985 (35)	2.022 (44)	0.435 (20)
-0.650	-3.98	2.329	0.321	0.645 (28)	3.618 (55)	2.231 (49)	0.515 (24)
13.432	-4.363	0.034	1.832	13.437 (60)	3.998 (56)	0.064 (5)	2.026 (45)
0.291	0.000	0.455	-0.698	0.295 (15)	0.365 (18)	0.358 (17)	0.504 (23)
-0.619	0.124	-0.361	1.108	0.615 (26)	0.490 (22)	0.459 (21)	1.302 (39)
0.105	-2.533	1.282	-0.447	0.110 (7)	2.167 (48)	1.184 (37)	0.254 (11)
0.158	-0.078	-0.868	0.091	0.162 (9)	0.288 (13)	0.966 (34)	0.285 (12)
-0.295	-3.477	2.230	0.758	0.290 (14)	3.111 (53)	2.133 (47)	0.951 (33)
-0.005	2.498	0.517	1.665	0.000(2.5)	2.863 (52)	0.419 (19)	1.859 (43)
-0.131	0.914	-0.533	1.865	0.126 (8)	1.280 (38)	0.630 (27)	2.059 (46)
2.434	-1.035	0.098	-1.616	2.438 (50)	0.669 (29)	0.000(2.5)	1.422 (41)
-3.144	-3.144	1.481	-0.901	3.139 (54)	2.778 (51)	1.383 (40)	0.707 (31)
-5.100	0.326	-0.777	-0.502	5.095 (57)	0.692 (30)	0.874 (32)	0.309 (16)

Table 9.2. *Averages and Variances for Thickness ADM and Thickness TR with Zero ADMs Deleted (Example 9.1).*

	ADM				TR			
	A	B	C	D	A	B	C	D
mean	2.680	1.703	1.020	0.987	1.021	0.989	0.714	0.702
variance	14.111	1.772	0.548	0.434	0.818	0.272	0.151	0.108

given in Table 9.1. Before testing for the homogeneity of variances one should examine the normality assumption so that the appropriate HOV test can be used. (Recall that the lack of normality can have serious effects on an HOV test that assumes normality.) Normality can be checked by constructing a normal probability plot of the thickness data for each of the four samples. A normal probability plot for nozzle A would indicate a significant departure from normality, and hence, a robust procedure should be used to test the HOV hypothesis. Table 9.2 contains the summary statistics for the ADM and the TR with the zero ADMs deleted. The decision lines for ANOMV-LEV (with $\alpha = 0.10$, df $= (56 - 4) = 52$, and the conservative critical value $h(0.10; 4, 40) = 2.26$) are

$$\text{UDL} = 2.671,$$
$$\text{LDL} = 0.523,$$

which results in rejection of the HOV hypothesis since the ADM mean for nozzle A of 2.680 (see Table 9.2) is higher than the upper decision line. The TR means all plots within the ANOMV-TR decision lines (see problem 3); hence, using this test the HOV hypothesis would not be rejected. In general ANOMV-TR is more robust than ANOMV-LEV (that is, it can be used in cases of extreme nonnormality); however, ANOMV-LEV is more powerful than ANOMV-TR.

9.2   ANOM-Type Randomization Tests

The idea behind a randomization test is to calculate a statistic based on the original samples and then repeatedly rearrange (or shuffle) the data (both within and between all the samples) and recalculate the same statistic for each rearrangment. If one assumes the samples come from populations that differ at most in the parameter being estimated by the sample statistic, then the distribution of the rearrangements can be used as a reference distribution to measure the unusualness of the original arrangement. The tests presented here will be based on permutation shuffling (observations assigned to treatments without replacement); however, the same tests can be performed using bootstrap shuffling (observations assigned to treatments with replacement).

RANDANOMV-R: A Randomization Test Version of ANOMV

Recall that in ANOMV the ratio of each of the treatment variances to the sum of the variances (over all treatments) is compared to a pair of critical values. If any of these ratios is higher than the upper critical value or lower than the lower critical value, the HOV hypothesis is rejected (which is equivalent to rejecting the HOV hypothesis whenever the maximum ratio is greater than the upper critical value or the minimum ratio is less than the lower critical value). Hence, the key statistics are these ratios. When normality holds, the joint probability distribution of these ratios is known and comparison to critical values is appropriate. Otherwise this is inappropriate (as illustrated in Example 4.9). In RANDANOMV-R, the randomization version of ANOMV, the variance ratios are calculated for the original sample and the maximum and minimum ratios are determined. That is,

$$AD_{max} = \max\left(\frac{S_i^2}{\sum_{i=1}^{I} S_i^2}\right),$$

$$AD_{min} = \min\left(\frac{S_i^2}{\sum_{i=1}^{I} S_i^2}\right).$$

After each (permutation) shuffle of the data the variance ratios as well as the maximum and minimum ratios are calculated, yielding

$$AD_{max}^q = \max\left(\frac{S_i^2}{\sum_{i=1}^{I} S_i^2}\right),$$

$$AD_{min}^q = \min\left(\frac{S_i^2}{\sum_{i=1}^{I} S_i^2}\right),$$

for $q = 1, \ldots, \mathcal{N}$. A decision chart for RANDANOMV-R with level of significance α has decision lines

$$UDL = \left(\sum_{i=1}^{I} S_i^2 \right) (AD_{max}^{(1-(\alpha/2))})$$

$$LDL = \left(\sum_{i=1}^{I} S_i^2 \right) (AD_{min}^{(\alpha/2)})$$,

where $AD_{max}^{(1-(\alpha/2))}$ is the $(1 - (\alpha/2))$ empirical quantile of the AD_{max}^q.

Example 9.2 (Deviation from target Thickness Data). Recall the deviation from target thickness data. The variances are $S_A^2 = 4.484^2$, $S_B^2 = 2.019^2$, $S_C^2 = 1.213^2$, and $S_D^2 = 1.137^2$, and the sum of the variances is 26.947. For $\mathcal{N} = 5000$ permutation shuffles and $\alpha = 0.05$, the RANDANOMV-R decision lines are

$$UDL = \left(\sum_{i=1}^{I} S_i^2 \right) (AD_{max}^{(1-(\alpha/2))}) = 26.947(0.722) = 19.46,$$

$$LDL = \left(\sum_{i=1}^{I} S_i^2 \right) (AD_{min}^{(\alpha/2)}) = 26.947(0.028) = 0.76.$$

The center line is $26.947/4 = 6.737$, the average of the sample variances. To determine decision lines against which the standard deviations can be plotted, one need only take square roots of the limits, as was done for the RANDANOMV-R chart in Figure 9.1. The decision lines make it clear that there are differences among the standard deviations. The standard deviation for nozzle A is significantly high (not desirable). This standard deviation is more than twice the size of the next largest value, which is probably of practical significance. One advantage of using this test rather than a test based on transformations is that the measurement units are preserved and comparisons of the standard deviations can be made directly. Additionally, one can determine empirical p-values for RANDANOMV-R by locating the quantile of the observed maximum and minimum variance ratios. In this example the maximum ratio (which is associated with nozzle A) is $4.484^2/26.947 = 0.746$. Since this was exceeded by only 44 of 5000 shuffles, one can assign an empirical p-value of $44/5000 = 0.0088$ to this test. This p-value should be compared to $\alpha/2$.

Appropriateness of and Drawbacks to RANDANOMV-R

No distributional assumptions are required for randomization tests. RANDANOMV-R is suitable for balanced designs. However, RANDANOMV-R should not be used when there are large differences among the average values for the I populations since the shuffling will be contaminated by the differences in the means. A direct solution to this problem is to center the data by subtracting a centering constant c_i from each observation (see step 1 in the next section) before performing RANDANOMV-R. A robust test results from letting the centering constant for sample i be the sample median. Note that the I sample variances of the centered observations are the same as the sample variances of the noncentered observations,

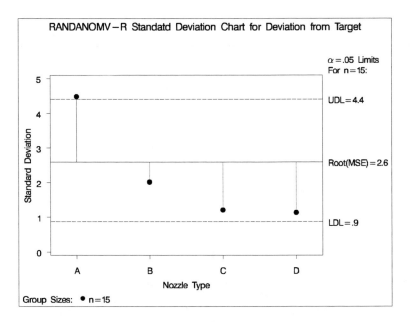

Figure 9.1. *RANDANOMV-R Chart for the Deviation from Target Thickness Data with Standard Deviations Plotted (Example 9.1).*

so the same points will be plotted on the RANDANOMV-R decision chart; that is, only the decision limits are affected by centering. See Bernard and Wludyka (2001) for a detailed discussion of alternatives when unequal location is suspected as well as Monte Carlo results showing that RANDANOMV-R is robust and powerful when compared to alternative robust tests.

Observe that the result in Example 9.2 arose from a single run of the randomization test. It is possible for a second performance (i.e., another 5000 samples) of the test to lead to a different conclusion. However, when the number of shuffles is large, the empirical p-values tend to be quite stable.

UBRANDANOMV-R: More General Randomization Test for Variances

In this section, a variance test applicable to unbalanced data will be presented. It has the added feature of including a centering constant that makes the test useful even in cases in which the populations being compared have different means. Note that n_i is the size of the sample from population i and $N = \sum_{i=1}^{I} n_i$ is the total sample size. Furthermore, there are K distinct sample sizes.

1. Given data y_{ij}, for each i determine centering constant c_i and let $x_{ij} = y_{ij} - c_i$.

2. Calculate the mean of each sample, $\bar{x}_i$.

3. Calculate $z_{ij} = x_{ij} - \bar{x}_i$.

4. Calculate $S_i^2 = \sum_{j=1}^{n_i} z_{ij}^2 / (n_i - 1)$ and

$$N_i = \frac{(n_i - 1)S_i^2}{\sum_{l=1}^{I}(n_l - 1)S_l^2}.$$

5. For the initial sample, calculate $AD_{max(k)} = max(N_i)$ over $i : n_i = n_k$ and $AD_{min(k)} = min(N_i)$ over $i : n_i = n_k$ for $k = 1, \ldots, K$. That is, calculate a min and max for each distinct sample size (using only those numerators associated with the sample size but using denominators based on the sum of all I variances).

6. Randomly shuffle the data x_{ij} some number of times, $\mathcal{N}$.

7. After each shuffle, calculate $S_i^2 = \sum_{j=1}^{n_i} z_{ij}^2 / (n_i - 1)$ and

$$N_i = \frac{(n_i - 1)S_i^2}{\sum_{l=1}^{I}(n_l - 1)S_l^2},$$

where $\bar{x}_i$ and $z_{ij} = x_{ij} - \bar{x}_i$ are based on the current shuffle.

8. Similar to step 5, calculate $AD_{max(k)}^q = max(N_i)$ over $i : n_i = n_k$ and $AD_{min(k)}^q = min(N_i)$ over $i : n_i = n_k$ for $k = 1, \ldots, K$.

For each k, the shuffling has produced a set of $\mathcal{N}$ $AD_{max(k)}^q$ labeled $q = 1, \ldots, \mathcal{N}$ as well as a set of $AD_{min(k)}^q$. The empirical quantiles of each set will be used to construct K sets of decision lines (one for each distinct sample size). A decision chart for UBRANDANOMV-R with level of significance α has decision lines

$$\boxed{\begin{aligned} UDL_k &= \left(\frac{\sum_{l=1}^{I}(n_l - 1)S_l^2}{n_i - 1} \right) (AD_{max(k)}^{(1-(\alpha/2K))}) \\ LDL_k &= \left(\frac{\sum_{l=1}^{I}(n_l - 1)S_l^2}{n_i - 1} \right) (AD_{min(k)}^{(\alpha/2K)}) \end{aligned}}$$

,

where $(AD_{max(k)}^{(1-(\alpha/2K))})$ is the $(1 - (\alpha/2K))$ empirical quantile of the $AD_{max(k)}^q$. The min is defined similarly. The center line is MS_e.

In the balanced case and with the centering constant $c_i = 0$, UBRANDANOMV-R is identical to RANDANOMV-R. The centering constant should be used when there is a strong suspicion that the populations under study have different means. Were the actual mean of each population known, that would be the ideal centering constant. Typical choices for the centering constant are the mean or median of sample i. The median centered version has been shown through a simulation study to have excellent robustness and power similar to ANOMV-LEV. In some cases it outperforms Levene's variance test (see Wludyka and Sa (2004)).

Example 9.3 (Time to Intubation Data). The data in Table 9.3 arose from a study of five presurgical anesthetics. Interest is in variability in the length of time for the anesthetic to become effective. The times and sample variances for each anesthetic are given in Table 9.3.

Table 9.3. *Time (Example 9.3).*

1 ($k = 1$)	2 ($k = 1$)	3 ($k = 2$)	4 ($k = 3$)		5 ($k = 3$)	
5.5	9.2	10.0	10.2	14.7	10.6	17.9
12.2	9.5	10.2	13.4	10.8	25.7	14.3
5.5	7.8	10.0	10.0	10.0	52.4	23.5
6.1	6.4	25.9	12.6	10.1	19.9	14.8
5.9	8.1	29.4	11.1	10.5	10.3	10.7
		14.9	10.1	11.6	10.2	10.1
		12.9	12.5	13.0	10.1	26.5
		39.0	10.2	13.5	11.8	14.7
		12.5	11.9	12.1	12.3	10.2
		10.2	10.0	12.3	17.9	10.7
$S_1^2 = 2.90^2$	$S_2^2 = 1.23^2$	$S_3^2 = 10.25^2$	$S_4^2 = 1.44^2$		$S_5^2 = 9.97^2$	
$n_1 = 5$	$n_2 = 5$	$n_3 = 10$	$n_4 = 20$		$n_5 = 20$	

Table 9.4. *UBRANDANOMV-R Decision Lines Summary (Example 9.3).*

Group	n_i	Admin	Admax	$\text{AD}_{\min(k)}^{(0.008)}$	$\text{AD}_{\max(k)}^{(0.993)}$	Variance	LDL	UDL
1	5	0.0021	0.0115	0.0005	0.5694	8.388	0.37	414.9
2	5	0.0021	0.0115	0.0005	0.5694	1.525	0.37	414.9
3	10	0.3247	0.3247	0.0052	0.6842	105.158	1.69	221.6
4	20	0.0134	0.6483	0.0177	0.8617	2.061	2.71	132.2
5	20	0.0134	0.6483	0.0177	0.8617	99.447	2.71	132.2

UBRANDANOMV-R was used to test the equality of the variances. Since there is evidence that the times to intubation are different with respect to location, the data were centered using the sample medians, which are 5.90, 8.10, 12.70, 11.35, and 13.30, respectively. Since there are three distinct sample sizes, $K = 3$, and $k = 1$ for $n = 5$, $k = 2$ for $n = 10$, and $k = 3$ for $n = 20$. From the initial sample one requires $\text{AD}_{\max(k)} = \max(N_i)$ over $i : n_i = n_k$ and $\text{AD}_{\min(k)} = \min(N_i)$ over $i : n_i = n_k$ for $k = 1, \ldots, K$. For $k = 1$ this yields

$$
\begin{aligned}
\text{AD}_{\max(1)} &= \max_{i=1,2}(N_i) \\
&= \frac{(5-1)(2.90)^2}{(5-1)(2.90)^2 + \cdots + (20-1)(9.97)^2} \\
&= \frac{33.55}{2914.73} = 0.01151.
\end{aligned}
$$

Two thousand shuffles were made ($\mathcal{N} = 2000$). Hence, three sets of 2000 $\text{AD}_{\max(k)}^q$ and $\text{AD}_{\min(k)}^q$ must be generated for each of $k = 1, 2, 3$. The decision lines are given in Table 9.4.

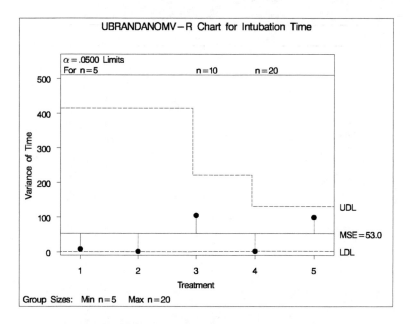

Figure 9.2. *UBRANDANOMV-R Chart for Intubation Time Data (Example 9.3).*

The calculations for $k = 1$ are

$$\text{UDL}_1 = \left(\frac{\sum_{l=1}^{I}(n_l - 1)S_l^2}{n_1 - 1}\right)(\text{AD}_{\max(k)}^{(1-(\alpha/2K))}) = (728.68)(0.5694) = 414.89,$$

$$\text{LDL}_1 = \left(\frac{\sum_{l=1}^{I}(n_l - 1)S_l^2}{n_1 - 1}\right)(\text{AD}_{\min(k)}^{(\alpha/2K)}) = (728.68)(0.000504) = 0.37.$$

From the UBRANDANOMV-R decision chart in Figure 9.2, one can see that the variance for treatment 4 is significantly low. (Equivalently, using Table 9.4, the lower decision line for sample size $n = 20$ is 2.71, which exceeds the variance of sample 4.)

Comparisons Among HOV Tests

Several HOV tests have been presented in this chapter and previously in Chapter 4. Each addresses the hypothesis

$$H_0 : \sigma_1^2 = \cdots = \sigma_I^2. \tag{9.1}$$

Note that given identical data, these tests will not always agree with respect to the decision regarding (9.1). The reason they differ is that the HOV hypothesis is being tested in somewhat different ways. Indeed, one might ask exactly what is meant by an HOV test. One could claim any set of steps leading to a decision regarding (9.1) as being such a test; however, a particular test will be useful in a given circumstance when the type I error rate is near the advertised level (α) and the test has power comparable to (or higher than) appropriate alternatives. ANOMV, RANDANOMV-R, and UBRANDANOMV-R are direct tests on the variances in that the calculations are done on functions of the sample variances. Whenever the normality assumption is tenable ANOMV is preferred, since it will in general have greater power than the randomization tests (and its type I error rate has been derived mathematically). ANOMV has the property that the data set will always lead to the same decision given α, which some researchers will find comforting (whereas randomization tests may not). The tests ANOMV-LEV and ANOMV-TR (see Chapter 4 for details) are, like the randomization tests, intended for use with populations that are fat-tailed or skewed, since ANOMV has an unacceptably high type I error rate for these type of populations. ANOMV-LEV is an indirect test of (9.1) since it actually compares the average ADMs of the populations. In general, if the ADMs differ, then the variances will tend to differ, so a test based on the sample ADMs is likely to reflect population differences in the variances. ANOMV-TR is based on the ranks of the ADMs so it is also an indirect test. Since information is lost in the ranks transformation, ANOMV-TR is less powerful than the Levene version of ANOMV and can lead to a decision different from ANOMV-LEV; however, ANOMV-TR is more robust to extreme departures from normality. These power conclusions are all based on Monte Carlo simulation studies (e.g., Conover et al. (1981), Wludyka and Nelson (1997b), Wludyka and Nelson (1999), Bernard and Wludyka (2001), and Wludyka and Sa (2004)).

All the HOV tests presented can be performed using a decision chart, which offers easy communication of the result. ANOMV, RANDANOMV-R, and UBRANDANOMV-R have the added advantage of having the variances (or standard deviations) plotted on the decision chart. Most users are more familiar with these metrics than ADMs or transformed ranks of ADMs.

These HOV tests have been presented for one-way layouts; however, they can be employed in more complex designs, such as two-way layouts (in a manner analogous to ANOM). Examples of this can be seen in Wludyka and Nelson (1997).

9.3 Distribution-Free ANOM Techniques

In general, ANOM is fairly robust to departures from normality. For obvious nonnormality when one cannot transform the data to normality, one of the following procedures might be useful. For comparison we will examine the results when each one is used to analyze the data in Example 9.4.

Example 9.4 (Exponential Data). Three exponential populations with means $\mu_1 = 5$, $\mu_2 = 1$, and $\mu_3 = 1$ were sampled. The data are given in Table 9.5, and a normal probability plot of the residuals is shown in Figure 9.3. It is clear from the plot that the data are not normally distributed.

Rank Tests

Bakir (1989) introduced an ANOM procedure based on ranks (ANOMR) for the one-way (possibly unbalanced) layout. He assumed that the I populations being compared are continuous and differ at most in their location parameters. Each observed value Y_{ij} is replaced by its rank in the combined set of observations (call these R_{ij}) and the ANOM is performed on the ranks using critical values provided by Bakir. That is, the $R_{i\cdot}$ are plotted

Table 9.5. *Data From Three Exponential Populations (Example 9.4).*

	Population		
	1	2	3
	6.0609	2.0810	1.1794
	17.2851	1.2452	0.0394
	9.3210	1.6899	0.6620
	3.0377	1.7831	0.7759
$\overline{Y}_{i\bullet}$	8.93	1.70	0.664

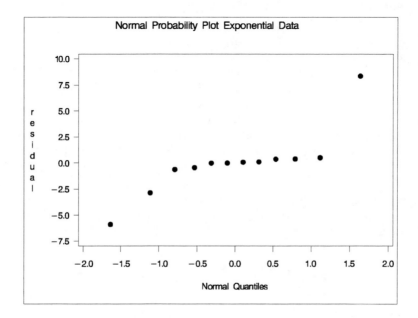

Figure 9.3. *Normal Probability of the Residuals for the Exponential Data (Example 9.4).*

with decision lines

$$\boxed{\overline{R}_{\bullet\bullet} \pm c(\alpha; I, n)} \,, \tag{9.2}$$

where $c(\alpha; I, n)$ is found in Bakir (1989). He provided exact critical values for $I = 3, 4$, various α, and combinations of small n_i. For equal sample sizes he suggests a Bonferonni approximation based on the Wilcoxon rank sum statistic, which was found to be satisfactory when α was chosen to coincide with the quantiles of the Wilcoxon rank sum statistic. This leads to α values such as 0.072.

Table 9.6. *Ranks and Treatment Rank Means (Example 9.5).*

	Population		
	1	2	3
	10	8	4
	12	5	1
	11	6	2
	9	7	3
$\overline{R}_{i\bullet}$	10.5	6.5	2.5

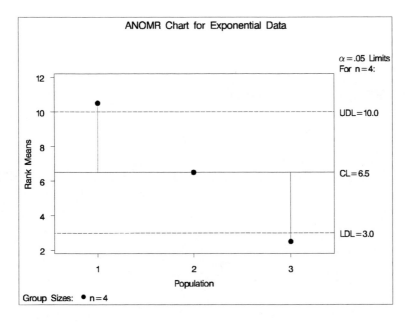

Figure 9.4. *ANOMR Chart (Example 9.5).*

For equal sample sizes and moderate to large samples, Bakir (1989) suggested an asymptotic procedure based on the fact that asymptotically the joint distribution of the $|\overline{R}_{i\bullet} - \overline{R}_{\bullet\bullet}|$ is the same as that of the $|\overline{Y}_{i\bullet} - \overline{Y}_{\bullet\bullet}|$ from the usual ANOM (i.e., when the Y_{ij} are independent normal with common variance). Thus, one could use the usual ANOM decision limits (equations (2.4)).

Example 9.5 (Exponential Data). The ranks and treatment rank means for these data are given in Table 9.6. The overall average rank $\overline{R}_{\bullet\bullet} = (N+1)/2 = 13/2 = 6.5, c(0.05; 3, 4) = 3.5$, and $c(0.01; 3, 4) = 4.25$. The ANOMR chart is shown in Figure 9.4 and suggest that population 1 has a significantly high mean and population 3 has a significantly low mean at $\alpha = 0.05$.

Table 9.7. *Ranks, Transformed Ranks, and Summary Statistics (Example 9.6).*

	Population				
1		2		3	
Rank	E_{1j}	Rank	E_{2j}	Rank	E_{3j}
10	1.2816	8	0.9154	4	0.4125
12	2.0537	5	0.5244	1	0.1004
11	1.5548	6	0.6433	2	0.2019
9	1.0803	7	0.7722	3	0.3055
$\overline{E}_{i\bullet}$	1.4926		0.7138		0.2551
s_i^2	0.1778		0.0283		0.0180

Transformed Ranks

Transforming the ranks R_{ij} using inverse normal scores leads to a procedure that does not require special critical values. Let

$$E_{ij} = \Phi^{-1}[0.5 + R_{ij}/(2N + 1)], \tag{9.3}$$

where $\Phi^{-1}(x)$ is the inverse of the standard normal distribution function. The usual ANOM procedure is then performed on the E_{ij}. More specifically, one computes the $\overline{E}_{i\bullet}$ and compares them with the decision lines

$$\boxed{\overline{E}_{\bullet\bullet} \pm h(\alpha; I, N - I)\sqrt{\mathrm{MS}_e}\sqrt{\frac{I-1}{N}}}, \tag{9.4}$$

where the MS_e is computed for the E_{ij}.

Example 9.6 (Exponential Data). For the data in Table 9.5, the ranks and the E_{ij} are given in Table 9.7. For example, Y_{12} is the largest value (i.e., has rank $R_{12} = 12$), and (equation (9.3))

$$E_{12} = \Phi^{-1}(0.5 + 12/25) = \Phi^{-1}(0.98) = 2.0537.$$

A normal probability plot of the residuals associated with the E_{ij} (Figure 9.5) suggests the transformation has been successful in establishing normality.

From Table 9.7 one can compute

$$\overline{E}_{\bullet\bullet} = \frac{1.492 + 0.7138 + 0.2551}{3} = 0.820,$$

$$\mathrm{MS}_e = \frac{0.1778 + 0.0283 + 0.0180}{3} = 0.0747.$$

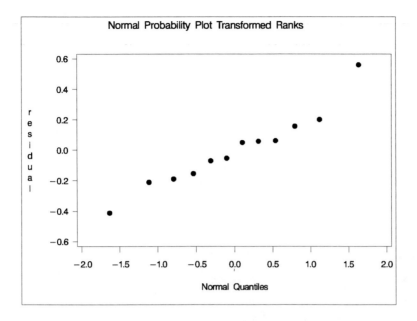

Figure 9.5. *Normal Probability of the Residuals for the E_{ij} (Example 9.6).*

For the ANOM with transformed ranks (ANOMTR) the decision lines (9.4) are

$$0.820 \pm h(0.01; 3, 9)\sqrt{0.0747}\sqrt{\frac{2}{12}}$$
$$\pm 3.84(0.1116)$$
$$\pm 0.429$$
$$(0.391, 1.249).$$

The $\alpha = 0.001$ decision lines are $(0.208, 1.433)$ (see problem 5). From the ANOMTR chart in Figure 9.6, one finds that population 1 has average values that are significantly high at the $\alpha = 0.01$ level, and population 3 has average values that are significantly low at the $\alpha = 0.01$ level.

A Randomization Test

Using randomly chosen permutations as the rearrangments, one can perform the permutation version of ANOM(PANOM) in two ways: one with symmetric decision lines and one with asymmetric decision lines. The symmetric version is performed as follows. First, compute

$$D_{\max} = \max_i |\overline{Y}_{i\bullet} - \overline{Y}_{\bullet\bullet}|$$

for the original data arrangement. Then for each $q = 1, \ldots, \mathcal{N}$ randomly chosen permutations, compute

$$D_{\max}^{(q)} = \max_i |\overline{Y}_{i(q)\bullet} - \overline{Y}_{\bullet\bullet}|.$$

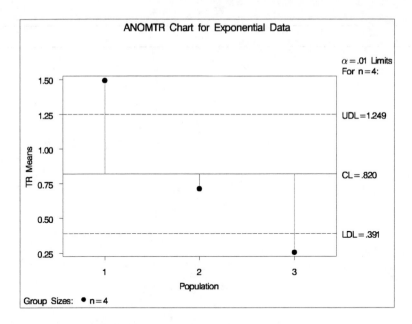

Figure 9.6. *ANOMTR Chart with $\alpha = 0.01$ Decision Lines (Example 9.6).*

Decision lines for the PANOM chart are

$$\boxed{\overline{Y}_{\bullet\bullet} \pm \mathcal{P}_\alpha}\ , \tag{9.5}$$

where $\mathcal{P}_\alpha$ is the upper α quantile of the randomization reference distribution of the $D_{\max}^{(q)}$.

Example 9.7 (Exponential Data). For the data in Table 9.5, critical values are $\mathcal{P}_{0.01} = 4.92$ and $\mathcal{P}_{0.001} = 5.16$ based on 10,000 permutations. The PANOM chart is given in Figure 9.7 and suggests that population 1 having a significantly high average at the $\alpha = 0.01$ level is the only significant effect.

The second randomization method produces asymmetrical decision lines. For each of $q = 1, \dots, \mathcal{N}$ randomly chosen permutations, compute

$$D_{\max}^{(q)} = \max_i (\overline{Y}_{i(q)\bullet} - \overline{Y}_{\bullet\bullet})$$

and

$$D_{\min}^{(q)} = \min_i (\overline{Y}_{i(q)\bullet} - \overline{Y}_{\bullet\bullet}).$$

Then the upper decision line for the PANOM chart is

$$\boxed{\overline{Y}_{\bullet\bullet} + \mathcal{P}_{\alpha/2}^+}\ , \tag{9.6}$$

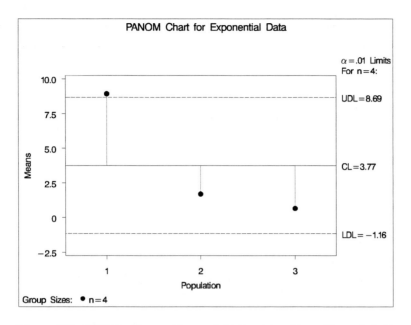

Figure 9.7. *PANOM Chart with $\alpha = 0.01$ Decision Lines (Example 9.7).*

where $\mathcal{P}_\alpha^+$ is the upper $\alpha/2$ quantile of the randomization reference distribution of the $D_{\max}^{(q)}$. The lower decision line for the PANOM chart is

$$\boxed{\overline{Y}_{\bullet\bullet} + \mathcal{P}_{\alpha/2}^-} \, , \qquad\qquad (9.7)$$

where $\mathcal{P}_\alpha^-$ is the lower $\alpha/2$ quantile of the randomization reference distribution of the $D_{\min}^{(q)}$.

Example 9.8 (Exponential Data). For the data in Table 9.5, critical values are $\mathcal{P}_{0.01}^+ = 4.924$ and $\mathcal{P}_{0.01}^- = -3.083$ based on 10,000 permutations. The PANOM chart is given in Figure 9.8 and suggests that population 1 has a significantly high average at the $\alpha = 0.02$ level and population 3 has a significantly low average (0.664 < LDL = 0.682).

Comparison of the Three Procedures

The previous examples cannot be used to systematically compare the three robust procedures. For that, see Nelson et al. (2005), in which it is shown that the procedures are robust (that is, they have type I error rates near the nominal rates for the distributions examined). However, the examples do illustrate that the three analyses of the exponential data give slightly different results. Using ANOMR and asymmetric PANOM, one finds population 1 has a significantly high mean and population 3 has a significantly low mean at the 0.05 level. The ANOMTR and PANOM both find population 1 to have a significantly high mean at (or close to) the 0.001 level. The ANOMTR also suggests that population 3 may have a significantly low mean. The ANOMTR does not require a special program for computation of critical values,

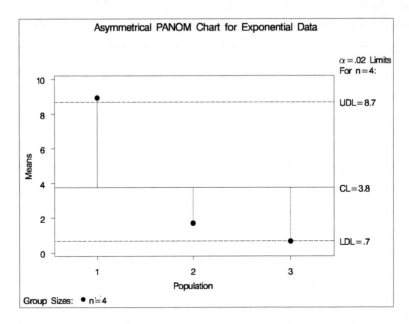

Figure 9.8. *Asymmetrical PANOM Chart with $\alpha = 0.02$ Decision Lines (Example 9.7).*

and, in addition, once the ranks have been transformed a normal probability plot can be used to assess the results before applying the ANOM.

Note on Randomization Tests

Repeated applications of a randomization test such as PANOM to the same data set (sample) can produce different conclusions (for details see, e.g., Nelson et al. (2005)). To see how this arises and to identify circumstances in which this is likely to occur, one can define the p-value for a randomization test such as PANOM in the following manner. Let Q be the number of distinct permutations of the sample. Of these, Q^* have the property that $D_{\max} > D_{\max}^{(q)}$. The p-value is then defined to be Q^*/Q, which is the proportion of the permutations for which the test statistic from the original sample exceeds a value from the permuted samples. An estimate of this p-value is given by $\mathcal{N}^*/\mathcal{N}$, which is the same proportion for the number of random permutations performed for the randomization test. For large $\mathcal{N}$ this estimate will tend to be near the true value. For a predetermined α, if the actual p-value is far from α, the probability of rejection of the hypothesis will be either nearly 0 or nearly 1; otherwise there is a substantial likelihood that repeated applications of the test will produce conflicting decisions.

Chapter 9 Problems

1. Transform the thickness data in Example 9.1 using the TR transformation to confirm the summary statistics in Table 9.2. Make a normal probability plot for the transformed data.

2. Perform the calculations to confirm the ANOMV-LEV decision lines for Example 9.1.

3. Perform the calculations to determine the ANOMV-TR decision lines for Example 9.1.

4. Determine the RANDANOMV-R decision lines for Example 4.7. Determine the empirical p-values.

5. Verify that the $\alpha = 0.001$ decision lines for the ANOMTR chart in Example 9.6 are correct.

Appendix A

Figures

A.1   ANOM Power Curves

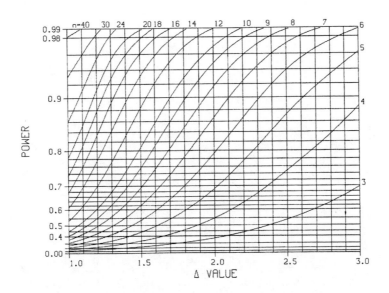

Figure A.1. *ANOM Power Curves for $\alpha = 0.05$ and $I = 3$.*

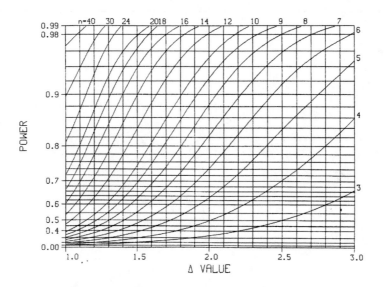

Figure A.2. *ANOM Power Curves for* $\alpha = 0.05$ *and* $I = 4$.

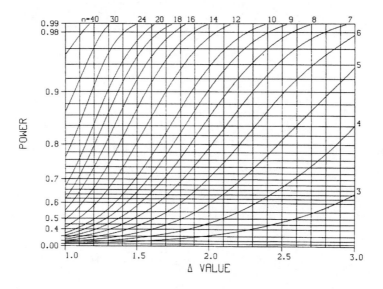

Figure A.3. *ANOM Power Curves for* $\alpha = 0.05$ *and* $I = 5$.

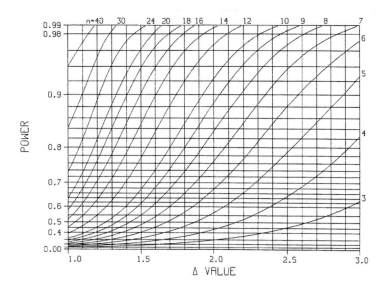

Figure A.4. *ANOM Power Curves for $\alpha = 0.05$ and $I = 6$.*

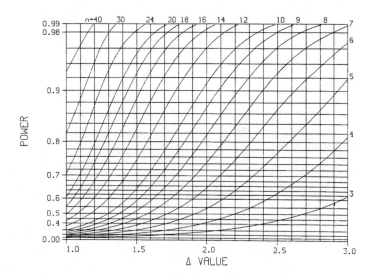

Figure A.5. *ANOM Power Curves for $\alpha = 0.05$ and $I = 7$.*

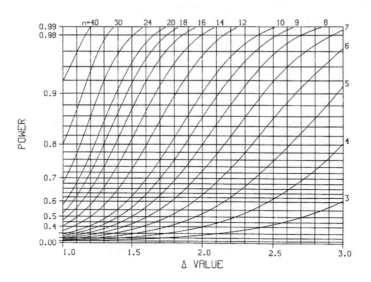

Figure A.6. *ANOM Power Curves for* $\alpha = 0.05$ *and* $I = 8$.

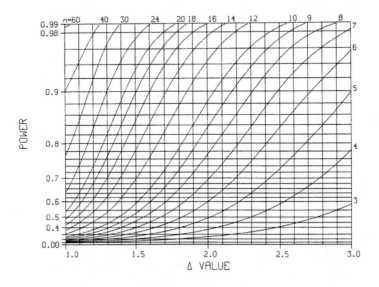

Figure A.7. *ANOM Power Curves for* $\alpha = 0.05$ *and* $I = 10$.

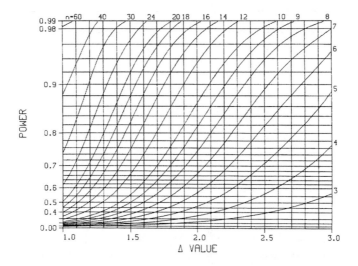

Figure A.8. *ANOM Power Curves for $\alpha = 0.05$ and $I = 12$.*

A.2   ANOMV Power Curves

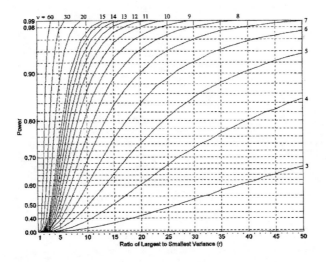

Figure A.9. *ANOMV Power Curves for $\alpha = 0.05$ and $I = 3$.*

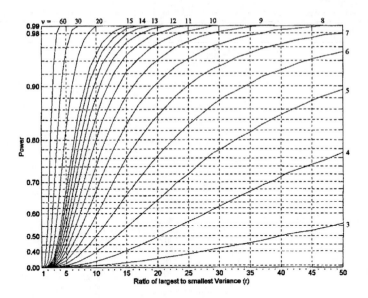

Figure A.10. *ANOMV Power Curves for* $\alpha = 0.05$ *and* $I = 4$.

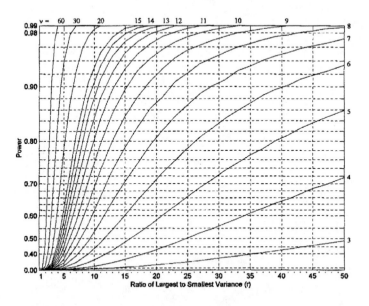

Figure A.11. *ANOMV Power Curves for* $\alpha = 0.05$ *and* $I = 5$.

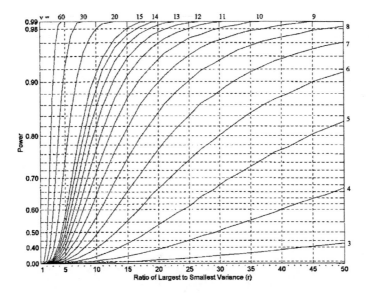

Figure A.12. *ANOMV Power Curves for $\alpha = 0.05$ and $I = 6$.*

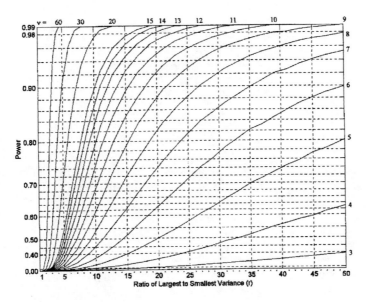

Figure A.13. *ANOMV Power Curves for $\alpha = 0.05$ and $I = 7$.*

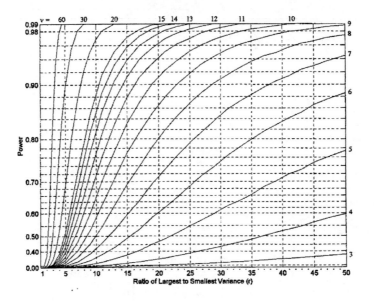

Figure A.14. *ANOMV Power Curves for* $\alpha = 0.05$ *and* $I = 8$.

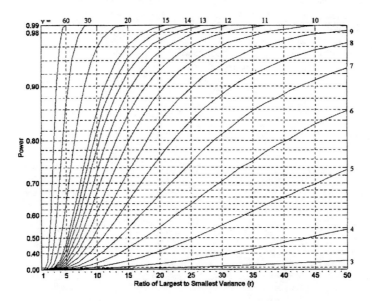

Figure A.15. *ANOMV Power Curves for* $\alpha = 0.05$ *and* $I = 10$.

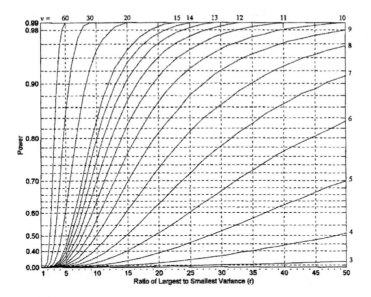

Figure A.16. *ANOMV Power Curves for* $\alpha = 0.05$ *and* $I = 12$.

A.3 HANOM Power Curves

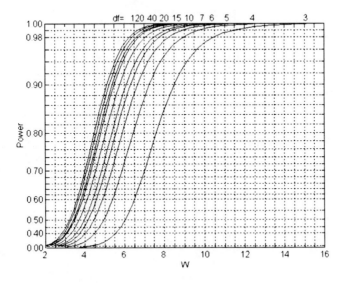

Figure A.17. *HANOM Power Curves for* $\alpha = 0.05$ *and* $I = 3$.

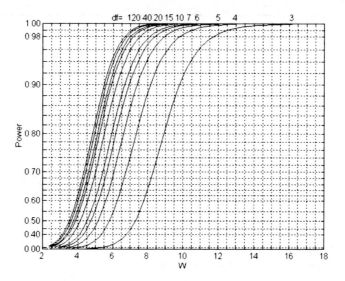

Figure A.18. *HANOM Power Curves for $\alpha = 0.05$ and $I = 4$.*

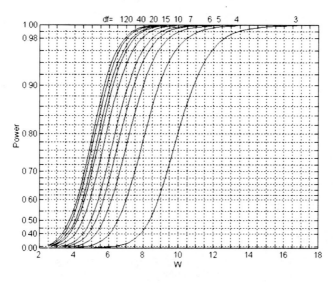

Figure A.19. *HANOM Power Curves for $\alpha = 0.05$ and $I = 5$.*

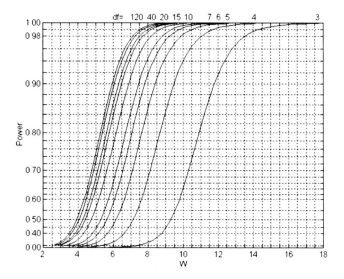

Figure A.20. *HANOM Power Curves for* $\alpha = 0.05$ *and* $I = 6$.

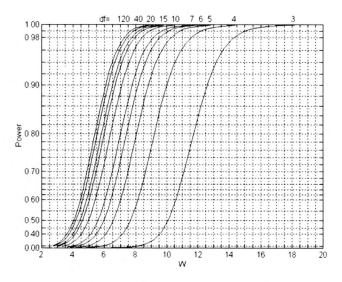

Figure A.21. *HANOM Power Curves for* $\alpha = 0.05$ *and* $I = 7$.

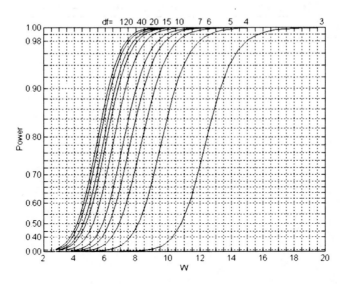

Figure A.22. *HANOM Power Curves for* $\alpha = 0.05$ *and* $I = 8$.

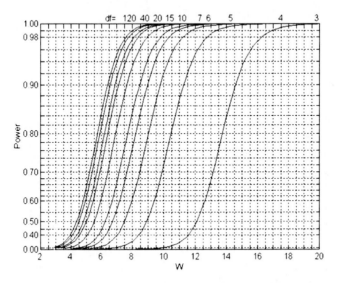

Figure A.23. *HANOM Power Curves for* $\alpha = 0.05$ *and* $I = 10$.

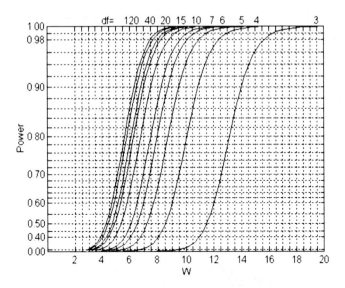

Figure A.24. *HANOM Power Curves for $\alpha = 0.05$ and $I = 12$.*

Appendix B
Tables

Table B.1. *Balanced ANOM Critical Values* $h(\alpha; k, \nu)$.

Level of Significance = 0.25

Number of Means Being Compared, k

ν	2	3	4	5	6	7	8	9	10	11	12	13	14	15	16	17	18	19	20
1	2.41	3.70	4.29	4.71	5.02	5.27	5.48	5.66	5.82	5.96	6.08	6.19	6.29	6.39	6.47	6.55	6.63	6.70	6.76
2	1.60	2.34	2.67	2.90	3.07	3.21	3.33	3.43	3.52	3.60	3.67	3.73	3.79	3.85	3.90	3.94	3.99	4.03	4.06
3	1.42	2.04	2.31	2.49	2.64	2.75	2.85	2.93	3.01	3.07	3.13	3.18	3.23	3.27	3.32	3.35	3.39	3.42	3.45
4	1.34	1.91	2.15	2.32	2.45	2.55	2.64	2.71	2.78	2.84	2.89	2.94	2.98	3.02	3.06	3.09	3.13	3.16	3.19
5	1.30	1.84	2.06	2.22	2.34	2.44	2.52	2.59	2.65	2.71	2.76	2.80	2.84	2.88	2.91	2.95	2.98	3.01	3.03
6	1.27	1.79	2.01	2.16	2.27	2.37	2.44	2.51	2.57	2.62	2.67	2.71	2.75	2.79	2.82	2.85	2.88	2.91	2.93
7	1.25	1.76	1.97	2.11	2.23	2.32	2.39	2.46	2.51	2.56	2.61	2.65	2.69	2.72	2.75	2.78	2.81	2.84	2.86
8	1.24	1.74	1.94	2.08	2.19	2.28	2.35	2.41	2.47	2.52	2.56	2.60	2.64	2.67	2.71	2.73	2.76	2.79	2.81
9	1.23	1.72	1.92	2.06	2.16	2.25	2.32	2.38	2.44	2.49	2.53	2.57	2.60	2.64	2.67	2.70	2.72	2.75	2.77
10	1.22	1.71	1.90	2.04	2.14	2.23	2.30	2.36	2.41	2.46	2.50	2.54	2.57	2.61	2.64	2.67	2.69	2.72	2.74
11	1.21	1.70	1.89	2.02	2.13	2.21	2.28	2.34	2.39	2.44	2.48	2.52	2.55	2.58	2.61	2.64	2.67	2.69	2.71
12	1.21	1.69	1.88	2.01	2.11	2.19	2.26	2.32	2.37	2.42	2.46	2.50	2.53	2.56	2.59	2.62	2.65	2.67	2.69
13	1.20	1.68	1.87	2.00	2.10	2.18	2.25	2.31	2.36	2.40	2.44	2.48	2.52	2.55	2.58	2.60	2.63	2.65	2.67
14	1.20	1.67	1.86	1.99	2.09	2.17	2.24	2.29	2.35	2.39	2.43	2.47	2.50	2.53	2.56	2.59	2.61	2.64	2.66
15	1.20	1.67	1.85	1.98	2.08	2.16	2.23	2.28	2.33	2.38	2.42	2.46	2.49	2.52	2.55	2.57	2.60	2.62	2.64
16	1.19	1.66	1.85	1.98	2.07	2.15	2.22	2.27	2.32	2.37	2.41	2.44	2.48	2.51	2.54	2.56	2.59	2.61	2.63
17	1.19	1.66	1.84	1.97	2.07	2.14	2.21	2.27	2.32	2.36	2.40	2.44	2.47	2.50	2.53	2.55	2.58	2.60	2.62
18	1.19	1.65	1.84	1.96	2.06	2.14	2.20	2.26	2.31	2.35	2.39	2.43	2.46	2.49	2.52	2.54	2.57	2.59	2.61
19	1.19	1.65	1.83	1.96	2.06	2.13	2.20	2.25	2.30	2.35	2.38	2.42	2.45	2.48	2.51	2.54	2.56	2.58	2.60
20	1.18	1.65	1.83	1.95	2.05	2.13	2.19	2.25	2.30	2.34	2.38	2.41	2.45	2.48	2.50	2.53	2.55	2.57	2.60
24	1.18	1.64	1.82	1.94	2.04	2.11	2.17	2.23	2.28	2.32	2.36	2.39	2.42	2.45	2.48	2.51	2.53	2.55	2.57
30	1.17	1.63	1.81	1.93	2.02	2.09	2.16	2.21	2.26	2.30	2.34	2.37	2.40	2.43	2.46	2.48	2.51	2.53	2.55
40	1.17	1.62	1.79	1.91	2.01	2.08	2.14	2.19	2.24	2.28	2.32	2.35	2.38	2.41	2.43	2.46	2.48	2.50	2.52
60	1.16	1.61	1.78	1.90	1.99	2.06	2.12	2.17	2.22	2.26	2.30	2.33	2.36	2.39	2.41	2.43	2.46	2.48	2.50
120	1.16	1.60	1.77	1.89	1.98	2.05	2.11	2.16	2.20	2.24	2.27	2.31	2.34	2.36	2.39	2.41	2.43	2.45	2.47
∞	1.15	1.59	1.76	1.87	1.96	2.03	2.09	2.14	2.18	2.22	2.25	2.29	2.31	2.34	2.36	2.39	2.41	2.43	2.45

Table B.1. (continued) *Balanced ANOM Critical Values $h(\alpha; k, \nu)$.*

Level of Significance = 0.1

ν	Number of Means Being Compared, k																		
	2	3	4	5	6	7	8	9	10	11	12	13	14	15	16	17	18	19	20
1	6.31	9.52	11.0	12.0	12.8	13.4	14.0	14.4	14.8	15.2	15.5	15.7	16.0	16.2	16.5	16.7	16.8	17.0	17.2
2	2.92	4.05	4.56	4.92	5.19	5.41	5.60	5.76	5.91	6.03	6.14	6.25	6.34	6.43	6.51	6.58	6.65	6.72	6.78
3	2.35	3.16	3.50	3.75	3.94	4.09	4.23	4.34	4.44	4.53	4.61	4.68	4.74	4.81	4.86	4.91	4.96	5.01	5.05
4	2.13	2.81	3.09	3.29	3.45	3.58	3.69	3.78	3.86	3.93	4.00	4.06	4.11	4.17	4.21	4.26	4.30	4.34	4.37
5	2.02	2.63	2.88	3.05	3.19	3.30	3.40	3.48	3.55	3.62	3.68	3.73	3.78	3.83	3.87	3.91	3.94	3.98	4.01
6	1.94	2.52	2.74	2.91	3.03	3.13	3.22	3.30	3.36	3.42	3.48	3.53	3.57	3.61	3.65	3.69	3.72	3.75	3.78
7	1.89	2.44	2.65	2.81	2.92	3.02	3.10	3.17	3.24	3.29	3.34	3.39	3.43	3.47	3.51	3.54	3.57	3.60	3.63
8	1.86	2.39	2.59	2.73	2.85	2.94	3.02	3.08	3.14	3.19	3.24	3.29	3.33	3.36	3.40	3.43	3.46	3.49	3.52
9	1.83	2.34	2.54	2.68	2.79	2.87	2.95	3.01	3.07	3.12	3.17	3.21	3.25	3.28	3.32	3.35	3.38	3.40	3.43
10	1.81	2.31	2.50	2.64	2.74	2.83	2.90	2.96	3.02	3.06	3.11	3.15	3.19	3.22	3.25	3.28	3.31	3.34	3.36
11	1.80	2.29	2.47	2.60	2.70	2.79	2.86	2.92	2.97	3.02	3.06	3.10	3.14	3.17	3.20	3.23	3.26	3.28	3.31
12	1.78	2.27	2.45	2.57	2.67	2.75	2.82	2.88	2.93	2.98	3.02	3.06	3.10	3.13	3.16	3.19	3.21	3.24	3.26
13	1.77	2.25	2.43	2.55	2.65	2.73	2.79	2.85	2.90	2.95	2.99	3.03	3.06	3.09	3.12	3.15	3.18	3.20	3.23
14	1.76	2.23	2.41	2.53	2.63	2.70	2.77	2.83	2.88	2.92	2.96	3.00	3.03	3.06	3.09	3.12	3.15	3.17	3.19
15	1.75	2.22	2.39	2.51	2.61	2.68	2.75	2.80	2.85	2.90	2.94	2.97	3.01	3.04	3.07	3.09	3.12	3.14	3.17
16	1.75	2.21	2.38	2.50	2.59	2.67	2.73	2.79	2.83	2.88	2.92	2.95	2.99	3.02	3.04	3.07	3.10	3.12	3.14
17	1.74	2.20	2.37	2.49	2.58	2.65	2.71	2.77	2.82	2.86	2.90	2.93	2.97	3.00	3.02	3.05	3.08	3.10	3.12
18	1.73	2.19	2.36	2.47	2.56	2.64	2.70	2.75	2.80	2.84	2.88	2.92	2.95	2.98	3.01	3.03	3.06	3.08	3.10
19	1.73	2.18	2.35	2.46	2.55	2.63	2.69	2.74	2.79	2.83	2.87	2.90	2.93	2.96	2.99	3.02	3.04	3.06	3.08
20	1.72	2.18	2.34	2.45	2.54	2.62	2.68	2.73	2.78	2.82	2.86	2.89	2.92	2.95	2.98	3.00	3.03	3.05	3.07
24	1.71	2.15	2.31	2.43	2.51	2.58	2.64	2.69	2.74	2.78	2.81	2.85	2.88	2.91	2.93	2.96	2.98	3.00	3.02
30	1.70	2.13	2.29	2.40	2.48	2.55	2.61	2.66	2.70	2.74	2.77	2.81	2.84	2.86	2.89	2.91	2.93	2.96	2.97
40	1.68	2.11	2.26	2.37	2.45	2.52	2.57	2.62	2.66	2.70	2.73	2.77	2.79	2.82	2.84	2.87	2.89	2.91	2.93
60	1.67	2.09	2.24	2.34	2.42	2.49	2.54	2.59	2.63	2.66	2.70	2.73	2.75	2.78	2.80	2.82	2.84	2.86	2.88
120	1.66	2.07	2.22	2.32	2.39	2.45	2.51	2.55	2.59	2.62	2.66	2.69	2.71	2.74	2.76	2.78	2.80	2.82	2.84
∞	1.65	2.05	2.19	2.29	2.36	2.42	2.47	2.52	2.55	2.59	2.62	2.65	2.67	2.69	2.72	2.74	2.75	2.77	2.79

Table B.1. (continued) *Balanced ANOM Critical Values* $h(\alpha; k, \nu)$.

Level of Significance = 0.05

ν	\multicolumn{19}{c}{Number of Means Being Compared, k}
	2	3	4	5	6	7	8	9	10	11	12	13	14	15	16	17	18	19	20
1	12.7	19.1	22.0	24.1	25.7	26.9	28.0	28.9	29.7	30.4	31.0	31.6	32.1	32.5	33.0	33.4	33.8	34.1	34.4
2	4.30	5.89	6.60	7.10	7.49	7.81	8.07	8.30	8.50	8.68	8.84	8.99	9.12	9.24	9.36	9.46	9.56	9.65	9.74
3	3.18	4.18	4.60	4.91	5.15	5.34	5.50	5.65	5.77	5.88	5.98	6.08	6.16	6.24	6.31	6.38	6.44	6.50	6.55
4	2.78	3.56	3.89	4.12	4.30	4.45	4.58	4.69	4.79	4.87	4.95	5.02	5.09	5.15	5.21	5.26	5.31	5.35	5.40
5	2.57	3.25	3.53	3.72	3.88	4.00	4.11	4.21	4.29	4.36	4.43	4.49	4.55	4.60	4.65	4.69	4.73	4.77	4.81
6	2.45	3.07	3.31	3.49	3.62	3.73	3.83	3.91	3.99	4.05	4.11	4.17	4.22	4.26	4.31	4.35	4.39	4.42	4.45
7	2.36	2.95	3.17	3.33	3.45	3.56	3.64	3.72	3.79	3.85	3.90	3.95	4.00	4.04	4.08	4.12	4.15	4.19	4.22
8	2.31	2.86	3.07	3.21	3.33	3.43	3.51	3.58	3.64	3.70	3.75	3.80	3.84	3.88	3.92	3.95	3.99	4.02	4.05
9	2.26	2.79	2.99	3.13	3.24	3.33	3.41	3.48	3.54	3.59	3.64	3.68	3.72	3.76	3.80	3.83	3.86	3.89	3.92
10	2.23	2.74	2.93	3.07	3.17	3.26	3.33	3.40	3.45	3.51	3.55	3.59	3.63	3.67	3.70	3.73	3.76	3.79	3.82
11	2.20	2.70	2.88	3.01	3.12	3.20	3.27	3.33	3.39	3.44	3.48	3.52	3.56	3.60	3.63	3.66	3.69	3.71	3.74
12	2.18	2.67	2.85	2.97	3.07	3.15	3.22	3.28	3.33	3.38	3.42	3.46	3.50	3.53	3.57	3.59	3.62	3.65	3.67
13	2.16	2.64	2.81	2.94	3.03	3.11	3.18	3.24	3.29	3.33	3.38	3.42	3.45	3.48	3.51	3.54	3.57	3.59	3.62
14	2.14	2.62	2.79	2.91	3.00	3.08	3.14	3.20	3.25	3.30	3.34	3.37	3.41	3.44	3.47	3.50	3.52	3.55	3.57
15	2.13	2.60	2.76	2.88	2.97	3.05	3.11	3.17	3.22	3.26	3.30	3.34	3.37	3.40	3.43	3.46	3.49	3.51	3.53
16	2.12	2.58	2.74	2.86	2.95	3.02	3.09	3.14	3.19	3.23	3.27	3.31	3.34	3.37	3.40	3.43	3.45	3.48	3.50
17	2.11	2.57	2.73	2.84	2.93	3.00	3.06	3.12	3.16	3.21	3.25	3.28	3.31	3.34	3.37	3.40	3.42	3.45	3.47
18	2.10	2.55	2.71	2.82	2.91	2.98	3.04	3.10	3.14	3.18	3.22	3.26	3.29	3.32	3.35	3.37	3.40	3.42	3.44
19	2.09	2.54	2.70	2.81	2.89	2.96	3.02	3.08	3.12	3.16	3.20	3.24	3.27	3.30	3.32	3.35	3.37	3.40	3.42
20	2.09	2.53	2.68	2.79	2.88	2.95	3.01	3.06	3.11	3.15	3.18	3.22	3.25	3.28	3.30	3.33	3.35	3.38	3.40
24	2.06	2.50	2.65	2.75	2.83	2.90	2.96	3.01	3.05	3.09	3.13	3.16	3.19	3.22	3.24	3.27	3.29	3.31	3.33
30	2.04	2.47	2.61	2.71	2.79	2.85	2.91	2.96	3.00	3.04	3.07	3.10	3.13	3.16	3.18	3.20	3.23	3.25	3.26
40	2.02	2.43	2.57	2.67	2.75	2.81	2.86	2.91	2.95	2.98	3.01	3.04	3.07	3.10	3.12	3.14	3.16	3.18	3.20
60	2.00	2.40	2.54	2.63	2.70	2.76	2.81	2.86	2.90	2.93	2.96	2.99	3.01	3.04	3.06	3.08	3.10	3.12	3.14
120	1.98	2.37	2.50	2.59	2.66	2.72	2.77	2.81	2.85	2.88	2.91	2.93	2.96	2.98	3.00	3.02	3.04	3.06	3.08
∞	1.96	2.34	2.47	2.56	2.62	2.68	2.72	2.76	2.80	2.83	2.86	2.88	2.90	2.93	2.95	2.97	2.98	3.00	3.02

Table B.1. (continued) *Balanced ANOM Critical Values* $h(\alpha; k, \nu)$.

Level of Significance = 0.01

Number of Means Being Compared, k

ν	2	3	4	5	6	7	8	9	10	11	12	13	14	15	16	17	18	19	20
1	63.7	95.7	110.	121.	129.	135.	140.	145.	149.	152.	155.	158.	160.	163.	165.	167.	169.	171.	172.
2	9.92	13.4	15.0	16.1	17.0	17.7	18.3	18.8	19.3	19.7	20.0	20.4	20.7	20.9	21.2	21.4	21.6	21.9	22.1
3	5.84	7.51	8.22	8.73	9.13	9.46	9.74	9.98	10.2	10.4	10.6	10.7	10.9	11.0	11.1	11.2	11.3	11.4	11.5
4	4.60	5.74	6.20	6.54	6.81	7.03	7.22	7.38	7.52	7.65	7.77	7.88	7.97	8.07	8.15	8.23	8.30	8.37	8.44
5	4.03	4.93	5.29	5.55	5.75	5.92	6.07	6.19	6.30	6.40	6.50	6.58	6.66	6.73	6.79	6.86	6.91	6.97	7.02
6	3.71	4.48	4.77	4.98	5.16	5.30	5.42	5.52	5.62	5.70	5.78	5.85	5.91	5.97	6.03	6.08	6.13	6.18	6.22
7	3.50	4.19	4.44	4.63	4.78	4.90	5.01	5.10	5.18	5.25	5.32	5.38	5.44	5.49	5.54	5.59	5.63	5.67	5.71
8	3.36	3.98	4.21	4.38	4.52	4.63	4.72	4.80	4.88	4.94	5.01	5.06	5.11	5.16	5.20	5.24	5.28	5.32	5.36
9	3.25	3.84	4.05	4.20	4.33	4.43	4.51	4.59	4.66	4.72	4.78	4.83	4.87	4.92	4.96	5.00	5.03	5.07	5.10
10	3.17	3.73	3.92	4.07	4.18	4.28	4.36	4.43	4.49	4.55	4.60	4.65	4.69	4.73	4.77	4.81	4.84	4.87	4.90
11	3.11	3.64	3.83	3.96	4.07	4.16	4.23	4.30	4.36	4.41	4.46	4.51	4.55	4.59	4.62	4.66	4.69	4.72	4.75
12	3.05	3.57	3.75	3.87	3.98	4.06	4.13	4.20	4.25	4.31	4.35	4.39	4.43	4.47	4.50	4.54	4.57	4.59	4.62
13	3.01	3.51	3.68	3.80	3.90	3.98	4.05	4.11	4.17	4.22	4.26	4.30	4.34	4.37	4.41	4.44	4.47	4.49	4.52
14	2.98	3.46	3.63	3.74	3.84	3.92	3.98	4.04	4.09	4.14	4.18	4.22	4.26	4.29	4.32	4.35	4.38	4.41	4.43
15	2.95	3.42	3.58	3.69	3.78	3.86	3.93	3.98	4.03	4.08	4.12	4.16	4.19	4.22	4.26	4.28	4.31	4.34	4.36
16	2.92	3.38	3.54	3.65	3.74	3.81	3.88	3.93	3.98	4.02	4.06	4.10	4.13	4.17	4.20	4.22	4.25	4.27	4.30
17	2.90	3.35	3.50	3.61	3.70	3.77	3.83	3.89	3.93	3.98	4.02	4.05	4.08	4.12	4.14	4.17	4.20	4.22	4.24
18	2.88	3.33	3.47	3.58	3.66	3.73	3.79	3.85	3.89	3.94	3.97	4.01	4.04	4.07	4.10	4.12	4.15	4.17	4.20
19	2.86	3.30	3.45	3.55	3.63	3.70	3.76	3.81	3.86	3.90	3.94	3.97	4.00	4.03	4.06	4.08	4.11	4.13	4.15
20	2.85	3.28	3.42	3.53	3.61	3.67	3.73	3.78	3.83	3.87	3.90	3.94	3.97	4.00	4.02	4.05	4.07	4.09	4.12
24	2.80	3.21	3.35	3.44	3.52	3.58	3.64	3.69	3.73	3.77	3.80	3.83	3.86	3.89	3.91	3.94	3.96	3.98	4.00
30	2.75	3.15	3.28	3.37	3.44	3.50	3.55	3.59	3.63	3.67	3.70	3.73	3.76	3.78	3.81	3.83	3.85	3.87	3.89
40	2.70	3.09	3.21	3.29	3.36	3.42	3.46	3.51	3.54	3.58	3.61	3.63	3.66	3.68	3.70	3.72	3.74	3.76	3.78
60	2.66	3.03	3.14	3.22	3.28	3.34	3.38	3.42	3.45	3.49	3.51	3.54	3.56	3.58	3.61	3.62	3.64	3.66	3.68
120	2.62	2.97	3.08	3.15	3.21	3.26	3.30	3.34	3.37	3.40	3.42	3.45	3.47	3.49	3.51	3.53	3.54	3.56	3.58
∞	2.58	2.91	3.01	3.08	3.14	3.19	3.22	3.26	3.29	3.32	3.34	3.36	3.38	3.40	3.42	3.44	3.45	3.47	3.48

Table B.1. (continued) *Balanced ANOM Critical Values* $h(\alpha; k, \nu)$.

Level of Significance = 0.001

ν	\multicolumn{19}{c}{Number of Means Being Compared, k}
	2	3	4	5	6	7	8	9	10	11	12	13	14	15	16	17	18	19	20
1	637.	957.	1103.	1207.	1285.	1349.	1401.	1446.	1485.	1520.	1551.	1579.	1605.	1628.	1650.	1670.	1689.	1707.	1724.
2	31.6	42.7	47.7	51.2	54.0	56.2	58.1	59.7	61.1	62.4	63.5	64.6	65.5	66.4	67.2	67.9	68.6	69.3	69.9
3	12.9	16.5	18.0	19.1	20.0	20.7	21.3	21.8	22.3	22.7	23.0	23.4	23.7	24.0	24.2	24.5	24.7	24.9	25.1
4	8.61	10.6	11.4	12.0	12.5	12.8	13.2	13.5	13.7	13.9	14.2	14.3	14.5	14.7	14.8	15.0	15.1	15.2	15.3
5	6.87	8.25	8.79	9.19	9.51	9.77	10.0	10.2	10.4	10.5	10.7	10.8	10.9	11.0	11.1	11.2	11.3	11.4	11.5
6	5.96	7.04	7.45	7.75	7.99	8.20	8.37	8.52	8.66	8.78	8.89	8.99	9.08	9.17	9.25	9.33	9.40	9.47	9.53
7	5.41	6.31	6.65	6.89	7.09	7.25	7.40	7.52	7.63	7.73	7.82	7.90	7.98	8.05	8.12	8.18	8.24	8.30	8.35
8	5.04	5.83	6.12	6.33	6.49	6.63	6.75	6.86	6.96	7.04	7.12	7.19	7.26	7.32	7.38	7.43	7.48	7.53	7.58
9	4.78	5.49	5.74	5.93	6.07	6.20	6.30	6.40	6.48	6.56	6.63	6.69	6.75	6.80	6.85	6.90	6.95	6.99	7.03
10	4.59	5.24	5.46	5.63	5.76	5.87	5.97	6.05	6.13	6.20	6.26	6.32	6.37	6.42	6.47	6.51	6.55	6.59	6.62
11	4.44	5.05	5.25	5.40	5.52	5.63	5.71	5.79	5.86	5.92	5.98	6.03	6.08	6.13	6.17	6.21	6.25	6.28	6.31
12	4.32	4.89	5.08	5.22	5.34	5.43	5.51	5.58	5.65	5.71	5.76	5.81	5.85	5.89	5.93	5.97	6.01	6.04	6.07
13	4.22	4.77	4.94	5.08	5.18	5.27	5.35	5.42	5.48	5.53	5.58	5.63	5.67	5.71	5.74	5.78	5.81	5.84	5.87
14	4.14	4.66	4.83	4.96	5.06	5.14	5.21	5.28	5.33	5.38	5.43	5.48	5.52	5.55	5.59	5.62	5.65	5.68	5.71
15	4.07	4.57	4.74	4.86	4.95	5.03	5.10	5.16	5.21	5.26	5.31	5.35	5.39	5.42	5.46	5.49	5.52	5.54	5.57
16	4.01	4.50	4.66	4.77	4.86	4.94	5.00	5.06	5.11	5.16	5.20	5.24	5.28	5.31	5.34	5.37	5.40	5.43	5.45
17	3.97	4.44	4.59	4.70	4.78	4.86	4.92	4.98	5.03	5.07	5.11	5.15	5.19	5.22	5.25	5.28	5.30	5.33	5.35
18	3.92	4.38	4.53	4.63	4.72	4.79	4.85	4.90	4.95	4.99	5.03	5.07	5.10	5.14	5.17	5.19	5.22	5.24	5.27
19	3.88	4.33	4.47	4.57	4.66	4.73	4.79	4.84	4.88	4.93	4.96	5.00	5.03	5.06	5.09	5.12	5.14	5.17	5.19
20	3.85	4.29	4.42	4.52	4.60	4.67	4.73	4.78	4.83	4.87	4.90	4.94	4.97	5.00	5.03	5.05	5.08	5.10	5.12
24	3.75	4.16	4.28	4.37	4.44	4.51	4.56	4.61	4.65	4.68	4.72	4.75	4.78	4.81	4.83	4.85	4.88	4.90	4.92
30	3.65	4.03	4.14	4.23	4.29	4.35	4.40	4.44	4.48	4.51	4.54	4.57	4.60	4.62	4.65	4.67	4.69	4.71	4.72
40	3.55	3.91	4.01	4.09	4.15	4.20	4.25	4.28	4.32	4.35	4.38	4.40	4.43	4.45	4.47	4.49	4.51	4.52	4.54
60	3.46	3.79	3.89	3.96	4.01	4.06	4.10	4.14	4.17	4.20	4.22	4.24	4.27	4.29	4.30	4.32	4.34	4.35	4.37
120	3.37	3.68	3.77	3.84	3.89	3.93	3.96	4.00	4.02	4.05	4.07	4.09	4.11	4.13	4.15	4.17	4.18	4.19	4.21
∞	3.29	3.58	3.66	3.72	3.76	3.80	3.84	3.86	3.89	3.91	3.93	3.95	3.97	3.99	4.00	4.02	4.03	4.04	4.06

Table B.2. *Sample Sizes for the ANOM, $\alpha = 0.1$.*

| | Δ = 3.00 | | | | | | | | Δ = 2.50 | | | | | | | | Δ = 2.00 | | | | | | | |
| | Power | | | | | | | | Power | | | | | | | | Power | | | | | | | |
k	.50	.60	.70	.75	.80	.90	.95	.99	.50	.60	.70	.75	.80	.90	.95	.99	.50	.60	.70	.75	.80	.90	.95	.99
3	2	3	3	3	3	4	4	6	3	3	3	4	4	5	6	7	3	4	4	5	5	7	8	11
4	2	3	3	3	3	4	5	6	3	3	4	4	4	5	6	8	4	4	5	5	6	7	9	12
5	2	3	3	3	3	4	5	6	3	3	4	4	5	6	7	9	4	5	5	6	6	8	9	13
6	3	3	3	3	4	4	5	7	3	3	4	4	5	6	7	9	4	5	5	6	7	8	10	13
7	3	3	3	3	4	5	5	7	3	4	4	4	5	6	7	9	4	5	6	6	7	9	10	14
8	3	3	3	4	4	5	5	7	3	4	4	5	5	6	7	9	4	5	6	6	7	9	11	14
9	3	3	3	4	4	5	5	7	3	4	4	5	5	6	7	10	4	5	6	7	7	9	11	14
10	3	3	3	4	4	5	6	7	3	4	4	5	5	6	7	10	4	5	6	7	7	9	11	15
11	3	3	3	4	4	5	6	7	3	4	5	5	5	7	8	10	5	6	7	7	8	10	11	15
12	3	3	3	4	4	5	6	7	3	4	5	5	5	7	8	10	5	6	7	7	8	10	12	15
13	3	3	4	4	4	5	6	7	3	4	5	5	5	7	8	10	5	6	7	7	8	10	12	15
14	3	3	4	4	4	5	6	7	3	4	5	5	6	7	8	10	5	6	7	7	8	10	12	16
15	3	3	4	4	4	5	6	8	3	4	5	5	6	7	8	10	5	6	7	8	8	10	12	16
16	3	3	4	4	4	5	6	8	4	4	5	5	6	7	8	11	5	6	7	8	8	10	12	16
17	3	3	4	4	4	5	6	8	4	4	5	5	6	7	8	11	5	6	7	8	8	11	12	16
18	3	3	4	4	4	5	6	8	4	4	5	5	6	7	8	11	5	6	7	8	9	11	13	16
19	3	3	4	4	4	5	6	8	4	4	5	5	6	7	8	11	5	6	7	8	9	11	13	17
20	3	3	4	4	4	5	6	8	4	4	5	6	6	7	9	11	5	6	8	8	9	11	13	17
24	3	3	4	4	4	5	6	8	4	4	5	6	6	7	9	11	5	6	8	8	9	11	13	17
30	3	3	4	4	5	6	6	8	4	5	5	6	6	8	9	12	6	7	8	9	9	12	14	18
40	3	4	4	4	5	6	7	9	4	5	6	6	7	8	9	12	6	7	8	9	10	12	14	18
60	3	4	4	5	5	6	7	9	5	5	6	7	7	9	10	13	7	8	9	10	11	13	15	20

Table B.2. (continued) *Sample Sizes for the ANOM, α = 0.1.*

| | Δ = 1.75 | | | | | | | | Δ = 1.50 | | | | | | | |
| | Power | | | | | | | | Power | | | | | | | |
k	.50	.60	.70	.75	.80	.90	.95	.99	.50	.60	.70	.75	.80	.90	.95	.99
3	4	5	5	6	6	8	10	14	5	6	7	7	8	11	13	18
4	4	5	6	7	7	9	11	15	5	6	8	8	9	12	15	20
5	4	5	6	7	8	10	12	16	6	7	8	9	10	13	16	21
6	5	6	7	7	8	10	13	17	6	7	9	10	11	14	17	23
7	5	6	7	8	9	11	13	17	6	8	9	10	11	14	17	23
8	5	6	7	8	9	11	13	18	7	8	10	11	12	15	18	24
9	5	6	8	8	9	12	14	19	7	8	10	11	12	15	19	25
10	5	7	8	9	9	12	14	19	7	9	10	11	13	16	19	25
11	6	7	8	9	10	12	15	19	7	9	11	12	13	16	19	26
12	6	7	8	9	10	12	15	20	7	9	11	12	13	17	20	26
13	6	7	8	9	10	13	15	20	8	9	11	12	13	17	20	27
14	6	7	9	9	10	13	15	20	8	9	11	12	14	17	20	27
15	6	7	9	9	11	13	16	20	8	10	12	13	14	18	21	27
16	6	7	9	10	11	13	16	21	8	10	12	14	14	18	21	28
17	6	8	9	10	11	13	16	21	8	10	12	13	14	18	21	28
18	6	8	9	10	11	14	16	21	8	10	12	13	15	18	22	28
19	6	8	9	10	11	14	16	21	8	10	12	13	15	18	22	29
20	7	8	9	10	11	14	16	21	9	10	12	14	15	19	22	29
24	7	8	10	11	12	14	17	22	9	11	13	14	15	19	23	30
30	7	9	10	11	12	15	18	23	10	11	13	15	16	20	24	31
40	8	9	11	12	13	16	18	24	10	12	14	16	17	21	25	32
60	8	10	12	13	14	17	20	25	11	13	16	17	19	23	27	34

Table B.2. (continued) *Sample Sizes for the ANOM, $\alpha = 0.1$.*

| | $\Delta = 1.25$ | | | | | | | | $\Delta = 1.00$ | | | | | | | |
| | Power | | | | | | | | Power | | | | | | | |
k	.50	.60	.70	.75	.80	.90	.95	.99	.50	.60	.70	.75	.80	.90	.95	.99
3	6	7	9	10	11	15	18	26	9	11	13	15	17	23	28	39
4	7	8	10	12	13	17	21	28	10	12	16	17	19	26	31	44
5	7	9	11	13	14	18	22	30	11	14	17	19	21	28	34	47
6	8	10	12	13	15	19	23	32	12	15	18	20	23	30	36	49
7	9	11	13	14	16	20	24	33	13	16	19	22	24	31	38	51
8	9	11	13	15	16	21	25	34	13	17	20	23	25	32	39	53
9	9	11	14	15	17	22	26	35	14	17	21	24	26	34	40	54
10	10	12	14	16	18	22	27	36	15	18	22	24	27	34	41	56
11	10	12	15	16	18	23	28	37	15	19	23	25	28	35	42	57
12	10	13	15	17	19	24	28	37	16	19	23	26	28	36	43	58
13	11	13	16	17	19	24	29	38	16	20	24	26	29	37	44	59
14	11	13	16	17	19	24	29	39	16	20	24	27	29	38	45	60
15	11	13	16	18	20	25	30	40	17	20	25	27	30	38	46	61
16	11	14	16	18	20	25	30	40	17	21	25	27	31	39	46	62
17	12	14	17	18	20	25	30	40	17	21	25	28	31	39	47	62
18	12	14	17	19	20	26	31	41	18	21	26	29	32	40	47	63
19	12	14	17	19	21	26	31	41	18	22	26	29	32	40	48	64
20	12	15	17	19	21	26	31	41	18	22	27	29	32	41	48	64
24	13	15	18	20	22	27	32	43	19	24	28	31	34	43	50	66
30	13	16	19	21	23	29	34	44	21	25	30	33	36	45	53	69
40	15	17	21	22	25	30	36	46	22	27	32	35	38	47	55	72
60	16	19	22	24	27	33	38	49	25	29	35	38	41	51	59	77

Table B.2. (continued) *Sample Sizes for the ANOM,* $\alpha = 0.05$.

k	Δ = 3.00 Power .50	.60	.70	.75	.80	.90	.95	.99	Δ = 2.50 Power .50	.60	.70	.75	.80	.90	.95	.99	Δ = 2.00 Power .50	.60	.70	.75	.80	.90	.95	.99
3	3	3	4	4	4	5	5	7	3	4	4	5	5	6	7	9	4	5	6	6	7	8	10	13
4	3	3	4	4	4	5	6	7	3	4	4	5	5	6	7	9	5	5	6	7	7	9	10	14
5	3	3	4	4	4	5	6	7	4	4	5	5	5	7	8	10	5	6	6	7	8	9	11	14
6	3	3	4	4	4	5	6	7	4	4	5	5	6	7	8	10	5	6	7	7	8	10	12	15
7	3	3	4	4	4	5	6	8	4	4	5	5	6	7	8	10	5	6	7	8	8	10	12	16
8	3	3	4	4	4	5	6	8	4	4	5	5	6	7	8	11	5	6	7	8	8	10	12	16
9	3	4	4	4	5	5	6	8	4	4	5	6	6	7	8	11	5	6	7	8	9	11	13	16
10	3	4	4	4	5	6	6	8	4	5	5	6	6	7	9	11	6	6	8	8	9	11	13	17
11	3	4	4	4	5	6	6	8	4	5	5	6	6	8	9	11	6	7	8	8	9	11	13	17
12	3	4	4	4	5	6	6	8	4	5	5	6	6	8	9	11	6	7	8	8	9	11	13	17
13	3	4	4	4	5	6	7	8	4	5	5	6	6	8	9	11	6	7	8	9	9	12	13	17
14	3	4	4	4	5	6	7	8	4	5	6	6	6	8	9	12	6	7	8	9	10	12	14	18
15	3	4	5	5	5	6	7	8	4	5	6	6	7	8	9	12	6	7	8	9	10	12	14	18
16	3	4	4	5	5	6	7	9	4	5	6	6	7	8	9	12	6	7	8	9	10	12	14	18
17	3	4	4	5	5	6	7	9	4	5	6	6	7	8	9	12	6	7	8	9	10	12	14	18
18	3	4	4	5	5	6	7	9	4	5	6	6	7	8	9	12	6	7	9	9	10	12	14	18
19	3	4	4	5	5	6	7	9	4	5	6	6	7	8	9	12	6	7	9	9	10	12	14	18
20	3	4	4	5	5	6	7	9	4	5	6	7	7	8	10	12	7	7	9	9	10	12	14	19
24	3	4	4	5	5	6	7	9	5	5	6	7	7	9	10	13	7	8	9	10	11	13	15	19
30	4	4	5	5	5	6	7	9	5	5	6	7	7	9	10	13	7	8	9	10	11	13	15	20
40	4	4	5	5	6	7	8	9	5	6	7	7	8	9	11	13	7	9	10	11	11	14	16	20
60	4	4	5	5	6	7	8	10	5	6	7	7	8	10	11	14	8	9	11	11	12	15	17	22

Table B.2. (continued) *Sample Sizes for the ANOM, α = 0.05.*

| | Δ = 1.75 | | | | | | | | Δ = 1.50 | | | | | | | |
| | Power | | | | | | | | Power | | | | | | | |
k	.50	.60	.70	.75	.80	.90	.95	.99	.50	.60	.70	.75	.80	.90	.95	.99
3	5	6	7	7	8	10	12	16	6	7	9	9	10	13	16	21
4	5	6	7	8	9	11	13	17	7	8	10	10	12	15	17	23
5	6	7	8	9	9	12	14	18	7	9	10	11	12	16	18	25
6	6	7	8	9	10	12	15	19	8	9	11	12	13	16	19	26
7	6	7	9	10	10	13	15	20	8	10	11	12	14	17	20	27
8	7	8	9	10	11	13	16	20	8	10	12	13	14	18	21	27
9	7	8	9	10	11	14	16	21	9	10	12	13	15	18	21	28
10	7	8	10	10	11	14	16	21	9	11	13	14	15	19	22	29
11	7	8	10	11	12	14	17	22	9	11	13	14	15	19	22	29
12	7	8	10	11	12	15	17	22	9	11	13	14	16	19	23	30
13	7	9	10	11	12	15	17	22	10	11	13	15	16	20	23	30
14	8	9	10	11	12	15	18	23	10	12	14	15	16	20	23	31
15	8	9	10	11	12	15	18	23	10	12	14	15	16	20	24	31
16	8	9	11	11	13	15	18	23	10	12	14	15	17	21	24	31
17	8	9	11	12	13	16	18	23	10	12	14	16	17	21	24	31
18	8	9	11	12	13	16	18	24	11	12	14	16	17	21	25	32
19	8	9	11	12	13	16	18	24	11	13	15	16	17	21	25	32
20	8	10	11	12	13	16	19	24	11	13	15	16	17	21	25	32
24	9	10	12	12	14	16	19	25	11	13	15	17	18	22	26	33
30	9	10	12	13	14	17	20	25	12	14	16	17	19	23	27	34
40	9	11	13	14	15	18	21	26	12	15	17	18	20	24	28	36
60	10	12	14	15	16	19	22	28	14	16	18	20	21	26	30	38

Table B.2. (continued) *Sample Sizes for the ANOM, $\alpha = 0.05$.*

| | Δ = 1.25 | | | | | | | | Δ = 1.00 | | | | | | | |
| | Power | | | | | | | | Power | | | | | | | |
k	.50	.60	.70	.75	.80	.90	.95	.99	.50	.60	.70	.75	.80	.90	.95	.99
3	8	10	12	13	14	18	22	30	12	14	17	19	21	27	33	45
4	9	11	13	14	16	20	24	33	13	16	20	22	24	31	37	50
5	10	12	14	16	17	22	26	35	15	18	21	24	26	33	40	54
6	11	13	15	17	18	23	27	37	16	19	23	25	28	35	42	56
7	11	13	16	17	19	24	29	38	17	20	24	27	29	37	44	58
8	12	14	17	18	20	25	30	39	18	21	25	28	30	38	45	60
9	12	14	17	19	21	26	30	40	18	22	26	29	31	39	47	62
10	12	15	18	19	21	26	31	41	19	23	27	29	32	40	48	63
11	13	15	18	20	22	27	32	42	20	23	28	30	33	41	49	64
12	13	16	19	20	22	27	32	42	20	24	28	31	34	42	50	66
13	14	16	19	21	23	28	33	43	21	24	29	31	34	43	51	67
14	14	16	19	21	23	28	33	43	21	25	29	32	35	44	52	67
15	14	17	20	21	23	29	34	44	21	25	30	33	36	45	52	68
16	14	17	20	22	24	29	34	45	22	26	31	33	36	45	53	69
17	15	17	20	22	24	30	35	45	22	26	31	34	37	46	54	70
18	15	17	20	22	24	30	35	45	22	27	32	34	37	46	54	71
19	15	18	21	22	24	30	35	46	22	27	32	35	38	47	55	71
20	15	18	21	23	25	31	36	46	23	28	32	35	38	47	55	72
24	16	18	22	24	26	32	37	48	24	29	34	37	40	49	57	74
30	17	20	23	25	27	33	38	49	26	30	35	38	42	51	59	77
40	18	21	24	26	28	35	40	51	28	32	38	41	44	54	62	80
60	20	23	26	28	30	37	43	54	30	35	41	44	47	57	66	84

Table B.2. (continued) *Sample Sizes for the ANOM, α = 0.01.*

	Δ = 3.00								Δ = 2.50								Δ = 2.00							
	Power								Power								Power							
k	.50	.60	.70	.75	.80	.90	.95	.99	.50	.60	.70	.75	.80	.90	.95	.99	.50	.60	.70	.75	.80	.90	.95	.99
3	4	4	5	5	5	6	7	9	5	5	6	6	7	8	9	11	6	7	8	9	9	11	13	16
4	4	5	5	5	6	7	7	9	5	6	6	7	7	8	10	12	7	8	9	9	10	12	14	18
5	4	5	5	5	6	7	8	9	5	6	7	7	7	9	10	13	7	8	9	10	11	13	15	18
6	4	5	5	5	6	7	8	9	5	6	7	7	8	9	10	13	7	8	10	10	11	13	15	19
7	4	5	5	6	6	7	8	10	5	6	7	7	8	9	11	13	8	9	10	10	11	13	15	20
8	4	5	5	6	6	7	8	10	5	6	7	7	8	9	11	13	8	9	10	11	12	14	16	20
9	4	5	5	6	6	7	8	10	6	6	7	8	8	10	11	14	8	9	10	11	12	14	16	20
10	4	5	5	6	6	7	8	10	6	6	7	8	8	10	11	14	8	9	11	11	12	14	16	21
11	4	5	5	6	6	7	8	10	6	6	7	8	8	10	11	14	8	9	11	11	12	14	17	21
12	4	5	5	6	6	7	8	10	6	7	7	8	8	10	11	14	8	9	11	12	12	15	17	21
13	4	5	6	6	6	7	8	10	6	7	7	8	8	10	11	14	8	10	11	12	12	15	17	22
14	4	5	6	6	6	7	8	10	6	7	7	8	8	10	11	14	8	10	11	12	13	15	17	22
15	4	5	6	6	6	7	8	10	6	7	7	8	9	10	12	14	9	10	11	12	13	15	17	22
16	4	5	6	6	6	7	8	10	6	7	8	8	9	10	12	15	9	10	11	12	13	15	18	22
17	5	5	6	6	6	7	8	10	6	7	8	8	9	10	12	15	9	10	11	12	13	16	18	22
18	5	5	6	6	6	7	8	11	6	7	8	8	9	10	12	15	9	10	11	12	13	16	18	22
19	5	5	6	6	6	8	9	11	6	7	8	8	9	10	12	15	9	10	12	12	13	16	18	23
20	5	5	6	6	6	8	9	11	6	7	8	8	9	11	12	15	9	11	12	13	13	16	18	23
24	5	5	6	6	7	8	9	11	6	7	8	8	9	11	12	15	9	11	12	13	14	16	19	23
30	5	5	6	6	7	8	9	11	6	7	8	9	9	11	13	16	10	11	12	13	14	17	19	24
40	5	5	6	7	7	8	9	11	7	8	9	9	10	11	13	16	10	11	13	14	15	17	20	25
60	5	6	6	7	7	8	10	12	7	8	9	10	10	12	14	17	11	12	14	15	16	18	21	26

Table B.2. (continued) *Sample Sizes for the ANOM, $\alpha = 0.01$.*

| | $\Delta = 1.75$ | | | | | | | | $\Delta = 1.50$ | | | | | | | |
| | Power | | | | | | | | Power | | | | | | | |
k	.50	.60	.70	.75	.80	.90	.95	.99	.50	.60	.70	.75	.80	.90	.95	.99
3	8	9	10	11	12	14	16	21	10	11	13	14	15	18	21	27
4	8	9	11	12	13	15	18	22	11	12	14	15	16	20	23	30
5	9	10	11	12	13	16	18	23	11	13	15	16	17	21	24	31
6	9	10	12	13	14	17	19	24	12	14	16	17	18	22	25	33
7	9	11	12	13	14	17	20	25	12	14	16	17	19	23	26	34
8	10	11	13	13	15	17	20	26	13	14	17	18	19	23	27	35
9	10	11	13	14	15	18	21	26	13	15	17	18	20	24	28	35
10	10	11	13	14	15	18	21	27	13	15	17	19	20	24	28	36
11	10	12	13	14	15	19	21	27	14	16	18	19	21	25	29	36
12	10	12	14	15	16	19	22	27	14	16	18	20	21	25	29	37
13	11	12	14	15	16	19	22	28	14	16	18	20	21	26	29	37
14	11	12	14	15	16	19	22	28	14	16	19	20	22	26	30	38
15	11	12	14	15	16	20	22	28	15	17	19	20	22	26	30	38
16	11	13	14	15	17	20	23	29	15	17	19	21	22	27	30	38
17	11	13	15	16	17	20	23	29	15	17	19	21	22	27	31	39
18	11	13	15	16	17	20	23	29	15	17	20	21	23	27	31	39
19	11	13	15	16	17	20	23	29	15	17	20	21	23	27	31	39
20	12	13	15	16	17	20	23	29	15	17	20	21	23	28	32	40
24	12	13	15	16	18	21	24	30	16	18	21	22	24	28	32	41
30	12	14	16	17	18	22	25	31	17	19	21	23	25	29	33	42
40	13	15	17	18	19	23	26	32	17	20	22	24	26	30	35	43
60	14	16	18	19	20	24	27	33	19	21	24	25	27	32	36	45

Table B.2. (continued) *Sample Sizes for the ANOM, α = 0.01.*

| | Δ = 1.25 | | | | | | | | Δ = 1.00 | | | | | | | |
| | Power | | | | | | | | Power | | | | | | | |
k	.50	.60	.70	.75	.80	.90	.95	.99	.50	.60	.70	.75	.80	.90	.95	.99
3	13	15	17	19	21	25	29	38	19	22	26	28	31	38	45	59
4	14	17	19	21	23	28	32	42	21	25	29	31	34	42	49	65
5	15	18	21	22	24	29	34	44	23	27	31	34	37	45	53	69
6	16	19	22	23	25	31	36	46	24	28	33	36	39	48	55	71
7	17	19	23	24	26	32	37	48	26	30	35	37	41	49	57	74
8	17	20	23	25	27	33	38	49	27	31	36	39	42	51	59	76
9	18	21	24	26	28	34	39	50	28	32	37	40	43	52	61	78
10	19	21	25	27	29	35	40	51	28	33	38	41	44	54	62	79
11	19	22	25	27	29	35	41	52	29	33	38	41	45	55	63	80
12	19	22	26	28	30	36	41	53	30	34	40	43	46	56	64	82
13	20	23	26	28	30	36	42	53	30	35	40	43	47	56	65	83
14	20	23	26	28	31	37	43	54	30	36	41	44	47	57	66	84
15	20	23	27	29	31	37	43	54	32	36	42	45	48	58	67	85
16	21	24	27	29	31	38	43	55	32	37	42	45	49	59	68	86
17	21	24	27	30	32	38	44	56	32	37	43	46	49	59	68	86
18	21	24	28	30	32	39	44	56	33	38	43	46	50	60	69	87
19	21	24	28	30	33	39	45	56	33	38	44	47	50	61	69	88
20	21	25	28	30	33	39	45	57	34	38	44	47	51	61	70	88
24	23	26	29	31	34	40	46	58	35	40	46	49	52	63	72	91
30	24	27	31	33	35	42	48	60	36	42	47	51	54	65	74	93
40	25	28	32	34	37	44	50	62	38	44	50	53	57	68	77	97
60	27	30	34	36	39	46	52	65	41	47	53	56	60	72	81	101

Table B.3. *Unbalanced ANOM Critical Values* $m(\alpha; k, \nu)$.

Level of Significance = 0.1

Number of Means Being Compared, k

ν	2	3	4	5	6	7	8	9	10	11	12	13	14	15	16	17	18	19	20
1	8.96	10.5	11.6	12.5	13.2	13.7	14.2	14.6	15.0	15.3	15.6	15.8	16.1	16.3	16.5	16.7	16.9	17.1	17.2
2	3.83	4.38	4.77	5.06	5.30	5.50	5.67	5.82	5.96	6.08	6.18	6.28	6.37	6.45	6.53	6.60	6.67	6.74	6.80
3	2.99	3.37	3.64	3.84	4.01	4.15	4.27	4.38	4.47	4.55	4.63	4.70	4.76	4.82	4.88	4.93	4.98	5.02	5.07
4	2.66	2.98	3.20	3.37	3.51	3.62	3.72	3.81	3.89	3.96	4.02	4.08	4.13	4.18	4.23	4.27	4.31	4.35	4.38
5	2.49	2.77	2.96	3.12	3.24	3.34	3.43	3.51	3.58	3.64	3.69	3.75	3.79	3.84	3.88	3.92	3.95	3.99	4.02
6	2.38	2.64	2.82	2.96	3.07	3.17	3.25	3.32	3.38	3.44	3.49	3.54	3.58	3.62	3.66	3.70	3.73	3.76	3.79
7	2.31	2.56	2.73	2.86	2.96	3.05	3.13	3.19	3.25	3.31	3.35	3.40	3.44	3.48	3.51	3.55	3.58	3.61	3.63
8	2.26	2.49	2.66	2.78	2.88	2.96	3.04	3.10	3.16	3.21	3.26	3.30	3.34	3.37	3.41	3.44	3.47	3.50	3.52
9	2.22	2.45	2.60	2.72	2.82	2.90	2.97	3.03	3.09	3.13	3.18	3.22	3.26	3.29	3.32	3.35	3.38	3.41	3.44
10	2.19	2.41	2.56	2.68	2.77	2.85	2.92	2.98	3.03	3.08	3.12	3.16	3.20	3.23	3.26	3.29	3.32	3.34	3.37
11	2.17	2.38	2.53	2.64	2.73	2.81	2.88	2.93	2.98	3.03	3.07	3.11	3.15	3.18	3.21	3.24	3.26	3.29	3.31
12	2.15	2.36	2.50	2.61	2.70	2.78	2.84	2.90	2.95	2.99	3.03	3.07	3.10	3.14	3.17	3.19	3.22	3.24	3.27
13	2.13	2.34	2.48	2.59	2.67	2.75	2.81	2.87	2.91	2.96	3.00	3.04	3.07	3.10	3.13	3.16	3.18	3.21	3.23
14	2.12	2.32	2.46	2.57	2.65	2.72	2.79	2.84	2.89	2.93	2.97	3.01	3.04	3.07	3.10	3.13	3.15	3.17	3.20
15	2.11	2.31	2.44	2.55	2.63	2.70	2.76	2.82	2.87	2.91	2.95	2.98	3.01	3.04	3.07	3.10	3.12	3.15	3.17
16	2.10	2.29	2.43	2.53	2.62	2.69	2.75	2.80	2.85	2.89	2.93	2.96	2.99	3.02	3.05	3.08	3.10	3.12	3.15
17	2.09	2.28	2.42	2.52	2.60	2.67	2.73	2.78	2.83	2.87	2.91	2.94	2.97	3.00	3.03	3.06	3.08	3.10	3.12
18	2.08	2.27	2.41	2.51	2.59	2.66	2.72	2.77	2.81	2.85	2.89	2.92	2.96	2.99	3.01	3.04	3.06	3.08	3.10
19	2.07	2.26	2.40	2.50	2.58	2.64	2.70	2.75	2.80	2.84	2.88	2.91	2.94	2.97	3.00	3.02	3.04	3.07	3.09
20	2.07	2.26	2.39	2.49	2.57	2.63	2.69	2.74	2.79	2.83	2.86	2.90	2.93	2.96	2.98	3.01	3.03	3.05	3.07
24	2.05	2.23	2.36	2.46	2.53	2.60	2.66	2.70	2.75	2.79	2.82	2.85	2.88	2.91	2.94	2.96	2.98	3.01	3.03
30	2.03	2.21	2.33	2.43	2.50	2.57	2.62	2.67	2.71	2.75	2.78	2.81	2.84	2.87	2.89	2.92	2.94	2.96	2.98
40	2.01	2.18	2.30	2.40	2.47	2.53	2.58	2.63	2.67	2.71	2.74	2.77	2.80	2.82	2.85	2.87	2.89	2.91	2.93
60	1.99	2.16	2.28	2.37	2.44	2.50	2.55	2.59	2.63	2.67	2.70	2.73	2.76	2.78	2.80	2.83	2.85	2.87	2.88
120	1.97	2.14	2.25	2.34	2.41	2.47	2.52	2.56	2.60	2.63	2.66	2.69	2.72	2.74	2.76	2.78	2.80	2.82	2.84
∞	1.95	2.11	2.23	2.31	2.38	2.43	2.48	2.52	2.56	2.59	2.62	2.65	2.67	2.70	2.72	2.74	2.76	2.77	2.79

Table B.3. (continued) *Unbalanced ANOM Critical Values $m(\alpha; k, \nu)$.*

Level of Significance = 0.05

ν	\multicolumn{19}{c}{Number of Means Being Compared, k}
	2	3	4	5	6	7	8	9	10	11	12	13	14	15	16	17	18	19	20
1	18.0	21.1	23.4	25.0	26.4	27.5	28.5	29.3	30.0	30.6	31.3	31.8	32.3	32.7	33.1	33.5	33.9	34.2	34.5
2	5.57	6.34	6.89	7.31	7.65	7.93	8.17	8.38	8.57	8.74	8.89	9.03	9.16	9.28	9.39	9.49	9.59	9.68	9.77
3	3.96	4.43	4.76	5.02	5.23	5.41	5.56	5.69	5.81	5.92	6.01	6.10	6.18	6.26	6.33	6.39	6.45	6.51	6.57
4	3.38	3.74	4.00	4.20	4.37	4.50	4.62	4.72	4.82	4.90	4.97	5.04	5.11	5.17	5.22	5.27	5.32	5.37	5.41
5	3.09	3.40	3.62	3.79	3.93	4.04	4.14	4.23	4.31	4.38	4.45	4.51	4.56	4.61	4.66	4.70	4.74	4.78	4.82
6	2.92	3.19	3.39	3.54	3.66	3.77	3.86	3.94	4.01	4.07	4.13	4.18	4.23	4.28	4.32	4.36	4.39	4.43	4.46
7	2.80	3.06	3.24	3.38	3.49	3.59	3.67	3.74	3.80	3.86	3.92	3.96	4.01	4.05	4.09	4.13	4.16	4.19	4.22
8	2.72	2.96	3.13	3.26	3.36	3.45	3.53	3.60	3.66	3.71	3.76	3.81	3.85	3.89	3.93	3.96	3.99	4.02	4.05
9	2.66	2.89	3.05	3.17	3.27	3.36	3.43	3.49	3.55	3.60	3.65	3.69	3.73	3.77	3.80	3.84	3.87	3.90	3.92
10	2.61	2.83	2.98	3.10	3.20	3.28	3.35	3.41	3.47	3.52	3.56	3.60	3.64	3.68	3.71	3.74	3.77	3.80	3.82
11	2.57	2.78	2.93	3.05	3.14	3.22	3.29	3.35	3.40	3.45	3.49	3.53	3.57	3.60	3.63	3.66	3.69	3.72	3.74
12	2.54	2.75	2.89	3.00	3.09	3.17	3.24	3.29	3.34	3.39	3.43	3.47	3.51	3.54	3.57	3.60	3.63	3.65	3.68
13	2.51	2.72	2.86	2.97	3.06	3.13	3.19	3.25	3.30	3.34	3.39	3.42	3.46	3.49	3.52	3.55	3.57	3.60	3.62
14	2.49	2.69	2.83	2.94	3.02	3.09	3.16	3.21	3.26	3.30	3.34	3.38	3.41	3.45	3.48	3.50	3.53	3.55	3.58
15	2.47	2.67	2.81	2.91	2.99	3.06	3.13	3.18	3.23	3.27	3.31	3.35	3.38	3.41	3.44	3.46	3.49	3.51	3.54
16	2.46	2.65	2.78	2.89	2.97	3.04	3.10	3.15	3.20	3.24	3.28	3.31	3.35	3.38	3.40	3.43	3.46	3.48	3.50
17	2.44	2.63	2.77	2.87	2.95	3.02	3.08	3.13	3.17	3.21	3.25	3.29	3.32	3.35	3.38	3.40	3.43	3.45	3.47
18	2.43	2.62	2.75	2.85	2.93	3.00	3.05	3.11	3.15	3.19	3.23	3.26	3.29	3.32	3.35	3.38	3.40	3.42	3.44
19	2.42	2.61	2.73	2.83	2.91	2.98	3.04	3.09	3.13	3.17	3.21	3.24	3.27	3.30	3.33	3.35	3.38	3.40	3.42
20	2.41	2.59	2.72	2.82	2.90	2.96	3.02	3.07	3.11	3.15	3.19	3.22	3.25	3.28	3.31	3.33	3.36	3.38	3.40
24	2.38	2.56	2.68	2.77	2.85	2.91	2.97	3.02	3.06	3.10	3.13	3.16	3.19	3.22	3.25	3.27	3.29	3.31	3.33
30	2.35	2.52	2.64	2.73	2.80	2.87	2.92	2.96	3.00	3.04	3.07	3.11	3.13	3.16	3.18	3.21	3.23	3.25	3.27
40	2.32	2.49	2.60	2.69	2.76	2.82	2.87	2.91	2.95	2.99	3.02	3.05	3.08	3.10	3.12	3.14	3.17	3.18	3.20
60	2.29	2.45	2.56	2.65	2.72	2.77	2.82	2.86	2.90	2.93	2.96	2.99	3.02	3.04	3.06	3.08	3.10	3.12	3.14
120	2.26	2.42	2.53	2.61	2.67	2.73	2.77	2.81	2.85	2.88	2.91	2.94	2.96	2.98	3.01	3.02	3.04	3.06	3.08
∞	2.24	2.39	2.49	2.57	2.63	2.68	2.73	2.77	2.80	2.83	2.86	2.88	2.91	2.93	2.95	2.97	2.98	3.00	3.02

Table B.3. (continued) *Unbalanced ANOM Critical Values* $m(\alpha; k, v)$.

Level of Significance = 0.01

v	\multicolumn{19}{c}{Number of Means Being Compared, k}
	2	3	4	5	6	7	8	9	10	11	12	13	14	15	16	17	18	19	20
1	90.0	106.	117.	125.	132.	138.	142.	146.	150.	153.	156.	159.	161.	164.	166.	168.	170.	171.	173.
2	12.7	14.4	15.7	16.6	17.4	18.0	18.5	19.0	19.4	19.8	20.1	20.5	20.8	21.0	21.3	21.5	21.7	21.9	22.1
3	7.13	7.91	8.48	8.92	9.28	9.58	9.84	10.1	10.3	10.4	10.6	10.8	10.9	11.0	11.2	11.3	11.4	11.5	11.6
4	5.46	5.99	6.36	6.66	6.90	7.10	7.27	7.43	7.57	7.69	7.80	7.91	8.00	8.09	8.17	8.25	8.32	8.39	8.45
5	4.70	5.11	5.40	5.63	5.81	5.97	6.11	6.23	6.33	6.43	6.52	6.60	6.67	6.74	6.81	6.87	6.93	6.98	7.03
6	4.27	4.61	4.86	5.05	5.20	5.33	5.45	5.55	5.64	5.72	5.80	5.86	5.93	5.99	6.04	6.09	6.14	6.18	6.23
7	4.00	4.30	4.51	4.68	4.81	4.93	5.03	5.12	5.20	5.27	5.34	5.39	5.45	5.50	5.55	5.60	5.64	5.68	5.72
8	3.81	4.08	4.27	4.42	4.55	4.65	4.74	4.82	4.89	4.96	5.02	5.07	5.12	5.17	5.21	5.25	5.29	5.33	5.36
9	3.67	3.92	4.10	4.24	4.35	4.45	4.53	4.61	4.67	4.73	4.79	4.84	4.88	4.92	4.96	5.00	5.04	5.07	5.10
10	3.57	3.80	3.97	4.10	4.20	4.29	4.37	4.44	4.50	4.56	4.61	4.66	4.70	4.74	4.78	4.81	4.84	4.88	4.91
11	3.48	3.71	3.87	3.99	4.09	4.17	4.25	4.31	4.37	4.42	4.47	4.51	4.55	4.59	4.63	4.66	4.69	4.72	4.75
12	3.42	3.63	3.78	3.90	4.00	4.08	4.15	4.21	4.26	4.31	4.36	4.40	4.44	4.48	4.51	4.54	4.57	4.60	4.63
13	3.36	3.57	3.71	3.83	3.92	4.00	4.06	4.12	4.18	4.22	4.27	4.31	4.34	4.38	4.41	4.44	4.47	4.50	4.52
14	3.32	3.52	3.66	3.77	3.85	3.93	3.99	4.05	4.10	4.15	4.19	4.23	4.26	4.30	4.33	4.36	4.39	4.41	4.44
15	3.28	3.47	3.61	3.71	3.80	3.87	3.93	3.99	4.04	4.08	4.12	4.16	4.20	4.23	4.26	4.29	4.31	4.34	4.36
16	3.25	3.43	3.57	3.67	3.75	3.82	3.88	3.94	3.99	4.03	4.07	4.11	4.14	4.17	4.20	4.23	4.25	4.28	4.30
17	3.22	3.40	3.53	3.63	3.71	3.78	3.84	3.89	3.94	3.98	4.02	4.06	4.09	4.12	4.15	4.17	4.20	4.22	4.25
18	3.19	3.37	3.50	3.60	3.68	3.74	3.80	3.85	3.90	3.94	3.98	4.01	4.04	4.07	4.10	4.13	4.15	4.18	4.20
19	3.17	3.35	3.47	3.57	3.65	3.71	3.77	3.82	3.86	3.90	3.94	3.97	4.01	4.03	4.06	4.09	4.11	4.13	4.16
20	3.15	3.32	3.45	3.54	3.62	3.68	3.74	3.79	3.83	3.87	3.91	3.94	3.97	4.00	4.03	4.05	4.07	4.10	4.12
24	3.09	3.25	3.37	3.46	3.53	3.59	3.64	3.69	3.73	3.77	3.80	3.83	3.86	3.89	3.91	3.94	3.96	3.98	4.00
30	3.03	3.18	3.29	3.38	3.45	3.50	3.55	3.60	3.64	3.67	3.70	3.73	3.76	3.78	3.81	3.83	3.85	3.87	3.89
40	2.97	3.12	3.22	3.30	3.37	3.42	3.47	3.51	3.54	3.58	3.61	3.63	3.66	3.68	3.71	3.73	3.74	3.76	3.78
60	2.91	3.05	3.15	3.23	3.29	3.34	3.38	3.42	3.46	3.49	3.51	3.54	3.56	3.59	3.61	3.63	3.64	3.66	3.68
120	2.86	2.99	3.09	3.16	3.21	3.26	3.30	3.34	3.37	3.40	3.43	3.45	3.47	3.49	3.51	3.53	3.55	3.56	3.58
∞	2.81	2.93	3.02	3.09	3.14	3.19	3.23	3.26	3.29	3.32	3.34	3.36	3.38	3.40	3.42	3.44	3.45	3.47	3.48

Table B.3. (continued) *Unbalanced ANOM Critical Values* $m(\alpha; k, \nu)$.

Level of Significance = 0.001

Number of Means Being Compared, k

ν	2	3	4	5	6	7	8	9	10	11	12	13	14	15	16	17	18	19	20
1	900.	1058.	1169.	1253.	1320.	1375.	1423.	1464.	1501.	1533.	1562.	1589.	1614.	1636.	1657.	1677.	1695.	1712.	1728.
2	40.4	45.8	49.7	52.6	55.0	57.1	58.8	60.3	61.6	62.8	63.9	64.9	65.8	66.6	67.4	68.2	68.8	69.5	70.1
3	15.7	17.3	18.6	19.5	20.3	20.9	21.5	22.0	22.4	22.8	23.1	23.5	23.8	24.1	24.3	24.6	24.8	25.0	25.2
4	10.1	11.0	11.7	12.2	12.6	13.0	13.3	13.5	13.8	14.0	14.2	14.4	14.6	14.7	14.9	15.0	15.1	15.3	15.4
5	7.88	8.50	8.95	9.31	9.60	9.85	10.1	10.2	10.4	10.6	10.7	10.8	11.0	11.1	11.2	11.3	11.4	11.4	11.5
6	6.74	7.21	7.56	7.83	8.06	8.25	8.41	8.56	8.69	8.80	8.91	9.01	9.10	9.19	9.27	9.34	9.41	9.48	9.54
7	6.05	6.44	6.73	6.95	7.13	7.29	7.43	7.55	7.65	7.75	7.84	7.92	8.00	8.07	8.13	8.20	8.25	8.31	8.36
8	5.60	5.93	6.18	6.37	6.53	6.66	6.78	6.88	6.97	7.06	7.13	7.20	7.27	7.33	7.39	7.44	7.49	7.54	7.58
9	5.28	5.58	5.79	5.96	6.10	6.22	6.32	6.41	6.50	6.57	6.64	6.70	6.76	6.81	6.86	6.91	6.95	7.00	7.04
10	5.04	5.31	5.51	5.66	5.78	5.89	5.99	6.07	6.14	6.21	6.27	6.33	6.38	6.43	6.47	6.52	6.56	6.59	6.63
11	4.86	5.11	5.29	5.43	5.54	5.64	5.73	5.80	5.87	5.93	5.99	6.04	6.09	6.13	6.17	6.21	6.25	6.29	6.32
12	4.71	4.94	5.11	5.24	5.35	5.44	5.52	5.59	5.66	5.71	5.77	5.81	5.86	5.90	5.94	5.98	6.01	6.04	6.07
13	4.59	4.81	4.97	5.10	5.20	5.28	5.36	5.42	5.48	5.54	5.59	5.63	5.67	5.71	5.75	5.78	5.82	5.85	5.87
14	4.50	4.71	4.86	4.97	5.07	5.15	5.22	5.28	5.34	5.39	5.44	5.48	5.52	5.56	5.59	5.62	5.65	5.68	5.71
15	4.41	4.61	4.76	4.87	4.96	5.04	5.11	5.17	5.22	5.27	5.31	5.35	5.39	5.43	5.46	5.49	5.52	5.55	5.57
16	4.34	4.54	4.68	4.78	4.87	4.95	5.01	5.07	5.12	5.16	5.21	5.25	5.28	5.32	5.35	5.38	5.41	5.43	5.46
17	4.28	4.47	4.60	4.71	4.79	4.86	4.93	4.98	5.03	5.08	5.12	5.15	5.19	5.22	5.25	5.28	5.31	5.33	5.36
18	4.23	4.41	4.54	4.64	4.72	4.79	4.85	4.91	4.95	5.00	5.04	5.07	5.11	5.14	5.17	5.20	5.22	5.25	5.27
19	4.19	4.36	4.49	4.58	4.66	4.73	4.79	4.84	4.89	4.93	4.97	5.00	5.04	5.07	5.09	5.12	5.15	5.17	5.19
20	4.14	4.32	4.44	4.53	4.61	4.68	4.73	4.78	4.83	4.87	4.91	4.94	4.97	5.00	5.03	5.06	5.08	5.10	5.13
24	4.02	4.18	4.29	4.38	4.45	4.51	4.56	4.61	4.65	4.69	4.72	4.75	4.78	4.81	4.83	4.86	4.88	4.90	4.92
30	3.90	4.05	4.15	4.23	4.30	4.35	4.40	4.44	4.48	4.51	4.54	4.57	4.60	4.62	4.65	4.67	4.69	4.71	4.73
40	3.79	3.92	4.02	4.09	4.15	4.20	4.25	4.29	4.32	4.35	4.38	4.40	4.43	4.45	4.47	4.49	4.51	4.53	4.54
60	3.68	3.81	3.89	3.96	4.02	4.06	4.10	4.14	4.17	4.20	4.22	4.24	4.27	4.29	4.31	4.32	4.34	4.36	4.37
120	3.58	3.69	3.78	3.84	3.89	3.93	3.97	4.00	4.03	4.05	4.07	4.10	4.11	4.13	4.15	4.17	4.18	4.19	4.21
∞	3.48	3.59	3.66	3.72	3.76	3.80	3.84	3.87	3.89	3.91	3.93	3.95	3.97	3.99	4.00	4.02	4.03	4.04	4.06

Table B.4. *ANOMV Critical Values for* $\alpha = 0.10$.

						Number of Variances (k)					
	3			4		5		6		7	
ν	lower	upper		lower	upper	lower	upper	lower	upper	lower	upper
3	0.0277	0.7868	con	0.0145	0.6805	0.0094	0.5964	0.0066	0.5310	0.0050	0.4791
3			lib	0.0151	0.6764	0.0097	0.5924	0.0069	0.5280	0.0052	0.4767
4	0.0464	0.7346	con	0.0262	0.6255	0.0177	0.5424	0.0130	0.4794	0.0100	0.4301
4			lib	0.0270	0.6215	0.0182	0.5389	0.0133	0.4761	0.0103	0.4274
5	0.0633	0.6962	con	0.0373	0.5863	0.0259	0.5046	0.0194	0.4437	0.0153	0.3966
5			lib	0.0383	0.5827	0.0265	0.5014	0.0198	0.4409	0.0156	0.3941
6	0.0780	0.6666	con	0.0474	0.5568	0.0335	0.4767	0.0254	0.4173	0.0202	0.3719
6			lib	0.0485	0.5534	0.0342	0.4737	0.0259	0.4149	0.0206	0.3697
7	0.0908	0.6430	con	0.0564	0.5337	0.0404	0.4549	0.0310	0.3971	0.0248	0.3529
7			lib	0.0575	0.5304	0.0411	0.4520	0.0315	0.3947	0.0253	0.3509
8	0.1020	0.6236	con	0.0644	0.5149	0.0466	0.4373	0.0360	0.3808	0.0290	0.3377
8			lib	0.0656	0.5117	0.0473	0.4347	0.0365	0.3786	0.0295	0.3360
9	0.1118	0.6074	con	0.0715	0.4992	0.0521	0.4227	0.0405	0.3673	0.0328	0.3253
9			lib	0.0727	0.4963	0.0529	0.4203	0.0411	0.3653	0.0333	0.3235
10	0.1205	0.5935	con	0.0779	0.4860	0.0571	0.4105	0.0446	0.3560	0.0363	0.3148
10			lib	0.0791	0.4831	0.0580	0.4082	0.0452	0.3541	0.0368	0.3132
11	0.1282	0.5815	con	0.0837	0.4746	0.0617	0.4000	0.0484	0.3464	0.0395	0.3059
11			lib	0.0849	0.4718	0.0625	0.3978	0.0490	0.3445	0.0400	0.3044
12	0.1352	0.5710	con	0.0889	0.4646	0.0658	0.3909	0.0518	0.3380	0.0424	0.2981
12			lib	0.0901	0.4620	0.0667	0.3887	0.0524	0.3362	0.0429	0.2966
13	0.1415	0.5617	con	0.0937	0.4558	0.0696	0.3828	0.0549	0.3306	0.0451	0.2914
13			lib	0.0949	0.4532	0.0705	0.3808	0.0556	0.3289	0.0456	0.2900
14	0.1472	0.5534	con	0.0980	0.4480	0.0731	0.3756	0.0578	0.3241	0.0476	0.2854
14			lib	0.0993	0.4455	0.0740	0.3737	0.0585	0.3225	0.0481	0.2840
15	0.1524	0.5459	con	0.1021	0.4409	0.0763	0.3693	0.0605	0.3182	0.0499	0.2800
15			lib	0.1033	0.4386	0.0772	0.3674	0.0612	0.3166	0.0504	0.2787
16	0.1573	0.5391	con	0.1058	0.4346	0.0793	0.3635	0.0630	0.3129	0.0520	0.2751
16			lib	0.1070	0.4323	0.0802	0.3616	0.0637	0.3114	0.0525	0.2739
17	0.1617	0.5329	con	0.1092	0.4288	0.0821	0.3582	0.0653	0.3082	0.0540	0.2708
17			lib	0.1104	0.4266	0.0829	0.3565	0.0660	0.3067	0.0545	0.2695
18	0.1659	0.5272	con	0.1124	0.4235	0.0847	0.3534	0.0675	0.3038	0.0559	0.2668
18			lib	0.1136	0.4214	0.0855	0.3517	0.0681	0.3024	0.0564	0.2656
19	0.1697	0.5220	con	0.1154	0.4187	0.0871	0.3490	0.0695	0.2998	0.0576	0.2631
19			lib	0.1166	0.4165	0.0879	0.3474	0.0702	0.2984	0.0581	0.2620
20	0.1733	0.5172	con	0.1182	0.4142	0.0894	0.3450	0.0715	0.2961	0.0593	0.2598
20			lib	0.1194	0.4121	0.0902	0.3433	0.0721	0.2948	0.0598	0.2586
21	0.1767	0.5126	con	0.1208	0.4100	0.0915	0.3412	0.0733	0.2927	0.0608	0.2566
21			lib	0.1220	0.4080	0.0923	0.3397	0.0739	0.2914	0.0613	0.2556
22	0.1798	0.5085	con	0.1233	0.4062	0.0935	0.3378	0.0749	0.2896	0.0623	0.2538
22			lib	0.1244	0.4042	0.0944	0.3362	0.0756	0.2883	0.0628	0.2527
23	0.1828	0.5046	con	0.1256	0.4026	0.0955	0.3345	0.0766	0.2866	0.0637	0.2511
23			lib	0.1267	0.4007	0.0962	0.3330	0.0772	0.2854	0.0641	0.2501
24	0.1856	0.5009	con	0.1278	0.3992	0.0972	0.3315	0.0781	0.2839	0.0650	0.2486
24			lib	0.1290	0.3973	0.0980	0.3300	0.0787	0.2827	0.0655	0.2476
25	0.1882	0.4975	con	0.1299	0.3961	0.0990	0.3286	0.0795	0.2813	0.0662	0.2463
25			lib	0.1311	0.3942	0.0997	0.3272	0.0801	0.2801	0.0667	0.2452
26	0.1908	0.4942	con	0.1319	0.3931	0.1006	0.3260	0.0809	0.2789	0.0674	0.2440
26			lib	0.1330	0.3913	0.1014	0.3245	0.0815	0.2778	0.0679	0.2431
27	0.1931	0.4912	con	0.1338	0.3903	0.1021	0.3235	0.0822	0.2766	0.0686	0.2419
27			lib	0.1349	0.3885	0.1029	0.3220	0.0828	0.2755	0.0690	0.2410
28	0.1954	0.4883	con	0.1356	0.3876	0.1036	0.3210	0.0834	0.2745	0.0697	0.2400
28			lib	0.1367	0.3859	0.1044	0.3197	0.0840	0.2734	0.0701	0.2391
29	0.1976	0.4855	con	0.1373	0.3851	0.1050	0.3188	0.0847	0.2724	0.0707	0.2382
29			lib	0.1384	0.3834	0.1058	0.3175	0.0852	0.2714	0.0711	0.2373
30	0.1996	0.4830	con	0.1390	0.3827	0.1064	0.3167	0.0858	0.2705	0.0717	0.2364
30			lib	0.1400	0.3811	0.1071	0.3154	0.0864	0.2694	0.0721	0.2355
31	0.2016	0.4805	con	0.1405	0.3805	0.1077	0.3146	0.0869	0.2687	0.0726	0.2347
31			lib	0.1416	0.3788	0.1084	0.3133	0.0875	0.2676	0.0731	0.2339
32	0.2035	0.4781	con	0.1420	0.3783	0.1089	0.3127	0.0879	0.2670	0.0735	0.2332
32			lib	0.1431	0.3767	0.1096	0.3115	0.0885	0.2659	0.0740	0.2323
33	0.2053	0.4758	con	0.1434	0.3763	0.1101	0.3109	0.0889	0.2653	0.0744	0.2316
33			lib	0.1445	0.3747	0.1108	0.3097	0.0895	0.2643	0.0749	0.2308
34	0.2070	0.4737	con	0.1449	0.3743	0.1112	0.3091	0.0899	0.2637	0.0753	0.2302
34			lib	0.1459	0.3727	0.1120	0.3079	0.0905	0.2627	0.0757	0.2294

Table B.4. (continued) *ANOMV Critical Values for* $\alpha = 0.10$.

		\multicolumn{10}{c}{Number of Variances (k)}									
		\multicolumn{2}{c}{8}	\multicolumn{2}{c}{9}	\multicolumn{2}{c}{10}	\multicolumn{2}{c}{11}	\multicolumn{2}{c}{12}
ν		lower	upper	lower	upper	lower	upper	lower	upper	lower	upper
3	con	0.0039	0.4371	0.0032	0.4029	0.0027	0.3736	0.0022	0.3483	0.0019	0.3262
3	lib	0.0041	0.4350	0.0033	0.4004	0.0027	0.3707	0.0023	0.3460	0.0020	0.3253
4	con	0.0081	0.3908	0.0067	0.3583	0.0056	0.3309	0.0048	0.3079	0.0042	0.2878
4	lib	0.0083	0.3884	0.0068	0.3559	0.0058	0.3293	0.0050	0.3063	0.0043	0.2861
5	con	0.0124	0.3589	0.0104	0.3281	0.0089	0.3028	0.0077	0.2809	0.0068	0.2627
5	lib	0.0127	0.3567	0.0106	0.3261	0.0091	0.3010	0.0079	0.2795	0.0069	0.2608
6	con	0.0167	0.3358	0.0140	0.3064	0.0121	0.2823	0.0105	0.2615	0.0093	0.2438
6	lib	0.0170	0.3339	0.0143	0.3047	0.0123	0.2806	0.0107	0.2600	0.0095	0.2428
7	con	0.0206	0.3182	0.0175	0.2898	0.0151	0.2665	0.0132	0.2465	0.0117	0.2298
7	lib	0.0209	0.3163	0.0177	0.2882	0.0153	0.2649	0.0134	0.2453	0.0119	0.2285
8	con	0.0242	0.3040	0.0206	0.2766	0.0179	0.2539	0.0157	0.2348	0.0140	0.2187
8	lib	0.0245	0.3022	0.0209	0.2751	0.0181	0.2525	0.0159	0.2336	0.0142	0.2176
9	con	0.0275	0.2923	0.0235	0.2657	0.0204	0.2436	0.0180	0.2252	0.0161	0.2096
9	lib	0.0278	0.2907	0.0238	0.2644	0.0207	0.2425	0.0183	0.2241	0.0163	0.2086
10	con	0.0305	0.2826	0.0261	0.2565	0.0228	0.2352	0.0201	0.2172	0.0180	0.2019
10	lib	0.0309	0.2811	0.0264	0.2553	0.0231	0.2340	0.0204	0.2162	0.0182	0.2009
11	con	0.0332	0.2742	0.0285	0.2489	0.0249	0.2279	0.0221	0.2104	0.0198	0.1955
11	lib	0.0336	0.2729	0.0289	0.2477	0.0252	0.2268	0.0223	0.2094	0.0200	0.1945
12	con	0.0357	0.2671	0.0308	0.2422	0.0269	0.2217	0.0239	0.2045	0.0214	0.1899
12	lib	0.0362	0.2658	0.0311	0.2410	0.0272	0.2206	0.0242	0.2035	0.0217	0.1890
13	con	0.0381	0.2609	0.0328	0.2363	0.0288	0.2162	0.0256	0.1994	0.0230	0.1850
13	lib	0.0385	0.2596	0.0332	0.2353	0.0291	0.2152	0.0258	0.1984	0.0232	0.1842
14	con	0.0402	0.2553	0.0348	0.2311	0.0305	0.2113	0.0271	0.1948	0.0244	0.1808
14	lib	0.0406	0.2541	0.0351	0.2301	0.0308	0.2104	0.0274	0.1940	0.0246	0.1799
15	con	0.0422	0.2503	0.0365	0.2265	0.0321	0.2070	0.0286	0.1908	0.0257	0.1769
15	lib	0.0427	0.2491	0.0369	0.2255	0.0324	0.2061	0.0288	0.1899	0.0259	0.1762
16	con	0.0441	0.2458	0.0382	0.2224	0.0336	0.2032	0.0299	0.1871	0.0269	0.1735
16	lib	0.0445	0.2447	0.0385	0.2214	0.0339	0.2023	0.0302	0.1863	0.0272	0.1728
17	con	0.0459	0.2418	0.0397	0.2186	0.0350	0.1997	0.0312	0.1838	0.0281	0.1704
17	lib	0.0463	0.2407	0.0401	0.2177	0.0353	0.1988	0.0315	0.1830	0.0283	0.1697
18	con	0.0475	0.2381	0.0412	0.2152	0.0363	0.1965	0.0324	0.1809	0.0292	0.1676
18	lib	0.0479	0.2371	0.0415	0.2143	0.0366	0.1956	0.0326	0.1801	0.0294	0.1669
19	con	0.0490	0.2347	0.0426	0.2121	0.0375	0.1935	0.0335	0.1781	0.0302	0.1650
19	lib	0.0494	0.2338	0.0429	0.2112	0.0378	0.1927	0.0338	0.1773	0.0305	0.1643
20	con	0.0505	0.2316	0.0438	0.2092	0.0387	0.1909	0.0346	0.1755	0.0312	0.1627
20	lib	0.0509	0.2307	0.0442	0.2084	0.0390	0.1901	0.0348	0.1748	0.0314	0.1619
21	con	0.0518	0.2288	0.0451	0.2065	0.0398	0.1884	0.0356	0.1732	0.0321	0.1604
21	lib	0.0522	0.2278	0.0454	0.2057	0.0401	0.1876	0.0358	0.1725	0.0323	0.1598
22	con	0.0531	0.2261	0.0462	0.2041	0.0408	0.1860	0.0365	0.1711	0.0330	0.1584
22	lib	0.0535	0.2252	0.0465	0.2033	0.0411	0.1853	0.0368	0.1704	0.0332	0.1578
23	con	0.0543	0.2237	0.0473	0.2018	0.0418	0.1839	0.0374	0.1691	0.0338	0.1566
23	lib	0.0547	0.2228	0.0476	0.2010	0.0421	0.1832	0.0377	0.1684	0.0340	0.1559
24	con	0.0555	0.2213	0.0483	0.1997	0.0427	0.1820	0.0383	0.1672	0.0346	0.1548
24	lib	0.0559	0.2205	0.0487	0.1989	0.0430	0.1813	0.0385	0.1666	0.0348	0.1542
25	con	0.0566	0.2192	0.0493	0.1976	0.0436	0.1801	0.0391	0.1655	0.0353	0.1531
25	lib	0.0570	0.2183	0.0496	0.1969	0.0439	0.1794	0.0393	0.1648	0.0355	0.1526
26	con	0.0576	0.2172	0.0502	0.1958	0.0445	0.1783	0.0398	0.1639	0.0360	0.1516
26	lib	0.0581	0.2163	0.0506	0.1951	0.0448	0.1777	0.0401	0.1633	0.0363	0.1510
27	con	0.0586	0.2153	0.0512	0.1940	0.0453	0.1767	0.0406	0.1623	0.0367	0.1501
27	lib	0.0590	0.2144	0.0515	0.1933	0.0456	0.1760	0.0408	0.1617	0.0369	0.1497
28	con	0.0596	0.2134	0.0520	0.1923	0.0461	0.1751	0.0413	0.1609	0.0374	0.1488
28	lib	0.0600	0.2127	0.0523	0.1916	0.0464	0.1745	0.0415	0.1603	0.0376	0.1483
29	con	0.0605	0.2117	0.0528	0.1908	0.0468	0.1737	0.0420	0.1595	0.0380	0.1476
29	lib	0.0609	0.2110	0.0531	0.1901	0.0471	0.1731	0.0422	0.1589	0.0382	0.1470
30	con	0.0614	0.2102	0.0536	0.1893	0.0475	0.1723	0.0426	0.1582	0.0386	0.1463
30	lib	0.0618	0.2094	0.0539	0.1886	0.0478	0.1717	0.0428	0.1577	0.0388	0.1458
31	con	0.0623	0.2086	0.0543	0.1879	0.0482	0.1710	0.0432	0.1570	0.0392	0.1451
31	lib	0.0626	0.2078	0.0547	0.1872	0.0485	0.1704	0.0435	0.1564	0.0394	0.1447
32	con	0.0630	0.2072	0.0551	0.1866	0.0488	0.1698	0.0438	0.1558	0.0397	0.1440
32	lib	0.0634	0.2064	0.0554	0.1859	0.0491	0.1691	0.0441	0.1553	0.0399	0.1436
33	con	0.0638	0.2058	0.0558	0.1853	0.0495	0.1686	0.0444	0.1547	0.0402	0.1430
33	lib	0.0642	0.2051	0.0561	0.1846	0.0497	0.1680	0.0446	0.1542	0.0405	0.1425
34	con	0.0646	0.2045	0.0564	0.1841	0.0501	0.1674	0.0449	0.1537	0.0408	0.1420
34	lib	0.0649	0.2038	0.0568	0.1834	0.0503	0.1669	0.0452	0.1532	0.0410	0.1415

Table B.4. (continued) *ANOMV Critical Values for* $\alpha = 0.05$.

							Number of Variances (k)				
	3			4		5		6		7	
ν	lower	upper		lower	upper	lower	upper	lower	upper	lower	upper
3	0.0167	0.8347	con	0.0090	0.7291	0.0058	0.6426	0.0041	0.5749	0.0031	0.5200
3			lib	0.0092	0.7271	0.0059	0.6409	0.0042	0.5730	0.0032	0.5194
4	0.0315	0.7821	con	0.0181	0.6705	0.0123	0.5842	0.0090	0.5175	0.0070	0.4654
4			lib	0.0184	0.6684	0.0125	0.5824	0.0092	0.5164	0.0071	0.4642
5	0.0461	0.7420	con	0.0276	0.6281	0.0192	0.5427	0.0144	0.4781	0.0114	0.4278
5			lib	0.0280	0.6262	0.0195	0.5410	0.0146	0.4769	0.0115	0.4268
6	0.0595	0.7103	con	0.0366	0.5957	0.0259	0.5114	0.0197	0.4486	0.0157	0.4000
6			lib	0.0371	0.5938	0.0262	0.5099	0.0199	0.4474	0.0159	0.3992
7	0.0715	0.6847	con	0.0449	0.5700	0.0322	0.4870	0.0248	0.4258	0.0199	0.3789
7			lib	0.0454	0.5682	0.0326	0.4856	0.0250	0.4247	0.0201	0.3778
8	0.0823	0.6635	con	0.0525	0.5490	0.0380	0.4673	0.0294	0.4074	0.0238	0.3618
8			lib	0.0530	0.5473	0.0384	0.4661	0.0297	0.4064	0.0240	0.3609
9	0.0919	0.6456	con	0.0593	0.5316	0.0433	0.4510	0.0337	0.3923	0.0274	0.3477
9			lib	0.0599	0.5299	0.0437	0.4497	0.0340	0.3913	0.0276	0.3467
10	0.1005	0.6302	con	0.0656	0.5167	0.0482	0.4373	0.0377	0.3796	0.0307	0.3358
10			lib	0.0662	0.5151	0.0486	0.4360	0.0380	0.3786	0.0309	0.3350
11	0.1084	0.6168	con	0.0713	0.5039	0.0527	0.4254	0.0414	0.3686	0.0338	0.3257
11			lib	0.0719	0.5023	0.0531	0.4242	0.0416	0.3677	0.0340	0.3249
12	0.1155	0.6050	con	0.0765	0.4926	0.0568	0.4150	0.0447	0.3591	0.0367	0.3170
12			lib	0.0771	0.4912	0.0572	0.4139	0.0450	0.3582	0.0369	0.3162
13	0.1220	0.5945	con	0.0813	0.4827	0.0605	0.4060	0.0478	0.3508	0.0393	0.3092
13			lib	0.0820	0.4813	0.0610	0.4048	0.0481	0.3500	0.0395	0.3086
14	0.1279	0.5851	con	0.0857	0.4739	0.0640	0.3979	0.0507	0.3434	0.0418	0.3025
14			lib	0.0864	0.4725	0.0645	0.3968	0.0510	0.3426	0.0420	0.3018
15	0.1334	0.5766	con	0.0898	0.4660	0.0673	0.3906	0.0534	0.3368	0.0441	0.2964
15			lib	0.0905	0.4646	0.0677	0.3896	0.0537	0.3360	0.0443	0.2957
16	0.1384	0.5690	con	0.0936	0.4588	0.0703	0.3841	0.0559	0.3309	0.0462	0.2909
16			lib	0.0943	0.4575	0.0708	0.3831	0.0562	0.3301	0.0464	0.2903
17	0.1431	0.5619	con	0.0971	0.4523	0.0732	0.3781	0.0583	0.3255	0.0482	0.2860
17			lib	0.0978	0.4510	0.0736	0.3772	0.0586	0.3247	0.0484	0.2854
18	0.1475	0.5555	con	0.1005	0.4463	0.0758	0.3727	0.0605	0.3205	0.0501	0.2815
18			lib	0.1011	0.4451	0.0762	0.3718	0.0608	0.3198	0.0503	0.2809
19	0.1515	0.5496	con	0.1036	0.4408	0.0783	0.3678	0.0626	0.3160	0.0519	0.2774
19			lib	0.1042	0.4396	0.0787	0.3668	0.0629	0.3153	0.0521	0.2767
20	0.1554	0.5440	con	0.1065	0.4357	0.0806	0.3632	0.0645	0.3118	0.0536	0.2736
20			lib	0.1071	0.4345	0.0810	0.3623	0.0648	0.3111	0.0538	0.2730
21	0.1589	0.5390	con	0.1092	0.4310	0.0828	0.3590	0.0664	0.3080	0.0551	0.2701
21			lib	0.1098	0.4299	0.0832	0.3581	0.0667	0.3073	0.0554	0.2695
22	0.1623	0.5342	con	0.1118	0.4266	0.0849	0.3550	0.0681	0.3044	0.0567	0.2668
22			lib	0.1124	0.4255	0.0853	0.3542	0.0684	0.3037	0.0569	0.2663
23	0.1655	0.5297	con	0.1142	0.4226	0.0869	0.3513	0.0698	0.3011	0.0581	0.2638
23			lib	0.1149	0.4214	0.0873	0.3505	0.0701	0.3004	0.0583	0.2633
24	0.1685	0.5255	con	0.1166	0.4187	0.0888	0.3479	0.0713	0.2980	0.0594	0.2609
24			lib	0.1172	0.4177	0.0892	0.3471	0.0716	0.2974	0.0597	0.2604
25	0.1714	0.5216	con	0.1188	0.4151	0.0905	0.3447	0.0728	0.2951	0.0607	0.2583
25			lib	0.1194	0.4141	0.0910	0.3439	0.0731	0.2945	0.0609	0.2578
26	0.1741	0.5179	con	0.1208	0.4118	0.0923	0.3416	0.0742	0.2924	0.0619	0.2558
26			lib	0.1215	0.4107	0.0926	0.3409	0.0745	0.2918	0.0622	0.2553
27	0.1766	0.5144	con	0.1229	0.4086	0.0939	0.3388	0.0756	0.2898	0.0631	0.2535
27			lib	0.1234	0.4076	0.0943	0.3380	0.0759	0.2892	0.0633	0.2530
28	0.1791	0.5111	con	0.1248	0.4056	0.0954	0.3361	0.0769	0.2874	0.0642	0.2513
28			lib	0.1253	0.4046	0.0958	0.3353	0.0772	0.2868	0.0644	0.2508
29	0.1814	0.5080	con	0.1266	0.4028	0.0969	0.3336	0.0782	0.2851	0.0653	0.2492
29			lib	0.1272	0.4017	0.0973	0.3328	0.0785	0.2845	0.0655	0.2487
30	0.1837	0.5050	con	0.1283	0.4001	0.0983	0.3311	0.0794	0.2829	0.0663	0.2472
30			lib	0.1289	0.3991	0.0987	0.3304	0.0796	0.2824	0.0666	0.2467
31	0.1858	0.5022	con	0.1300	0.3975	0.0997	0.3289	0.0805	0.2808	0.0673	0.2453
31			lib	0.1306	0.3965	0.1001	0.3281	0.0808	0.2803	0.0675	0.2449
32	0.1878	0.4995	con	0.1316	0.3950	0.1010	0.3267	0.0816	0.2789	0.0683	0.2436
32			lib	0.1322	0.3941	0.1013	0.3260	0.0819	0.2783	0.0685	0.2431
33	0.1898	0.4969	con	0.1331	0.3927	0.1022	0.3246	0.0827	0.2770	0.0692	0.2419
33			lib	0.1337	0.3918	0.1026	0.3239	0.0829	0.2765	0.0694	0.2414
34	0.1917	0.4944	con	0.1346	0.3905	0.1034	0.3226	0.0837	0.2752	0.0701	0.2402
34			lib	0.1351	0.3896	0.1038	0.3219	0.0840	0.2747	0.0703	0.2398

Table B.4. (continued) *ANOMV Critical Values for* $\alpha = 0.05$.

		Number of Variances (k)									
		8		9		10		11		12	
ν		lower	upper	lower	upper	lower	upper	lower	upper	lower	upper
3	con	0.0025	0.4763	0.0020	0.4385	0.0017	0.4075	0.0014	0.3795	0.0012	0.3559
3	lib	0.0025	0.4742	0.0020	0.4368	0.0017	0.4063	0.0014	0.3783	0.0012	0.3549
4	con	0.0056	0.4231	0.0047	0.3887	0.0039	0.3590	0.0034	0.3343	0.0030	0.3129
4	lib	0.0057	0.4219	0.0047	0.3869	0.0040	0.3583	0.0034	0.3331	0.0030	0.3120
5	con	0.0093	0.3878	0.0078	0.3547	0.0066	0.3269	0.0058	0.3038	0.0051	0.2838
5	lib	0.0094	0.3868	0.0078	0.3537	0.0067	0.3262	0.0058	0.3027	0.0051	0.2827
6	con	0.0130	0.3616	0.0109	0.3300	0.0094	0.3038	0.0082	0.2815	0.0073	0.2629
6	lib	0.0131	0.3607	0.0110	0.3291	0.0095	0.3032	0.0083	0.2811	0.0073	0.2620
7	con	0.0165	0.3416	0.0140	0.3112	0.0121	0.2860	0.0106	0.2648	0.0094	0.2467
7	lib	0.0166	0.3406	0.0141	0.3103	0.0122	0.2854	0.0107	0.2642	0.0095	0.2461
8	con	0.0198	0.3254	0.0169	0.2962	0.0147	0.2721	0.0129	0.2515	0.0115	0.2341
8	lib	0.0200	0.3248	0.0170	0.2955	0.0148	0.2715	0.0130	0.2509	0.0116	0.2337
9	con	0.0229	0.3123	0.0196	0.2839	0.0171	0.2606	0.0151	0.2408	0.0135	0.2239
9	lib	0.0231	0.3117	0.0197	0.2833	0.0172	0.2599	0.0152	0.2403	0.0135	0.2234
10	con	0.0258	0.3015	0.0221	0.2737	0.0193	0.2508	0.0171	0.2317	0.0153	0.2154
10	lib	0.0260	0.3007	0.0223	0.2731	0.0194	0.2503	0.0172	0.2312	0.0154	0.2150
11	con	0.0285	0.2921	0.0245	0.2650	0.0214	0.2427	0.0190	0.2240	0.0170	0.2081
11	lib	0.0286	0.2914	0.0246	0.2643	0.0215	0.2421	0.0191	0.2235	0.0171	0.2077
12	con	0.0309	0.2839	0.0267	0.2574	0.0233	0.2356	0.0207	0.2172	0.0186	0.2018
12	lib	0.0311	0.2833	0.0268	0.2569	0.0235	0.2351	0.0208	0.2168	0.0187	0.2014
13	con	0.0332	0.2768	0.0287	0.2509	0.0252	0.2294	0.0224	0.2115	0.0201	0.1964
13	lib	0.0334	0.2762	0.0288	0.2502	0.0253	0.2290	0.0225	0.2111	0.0202	0.1960
14	con	0.0354	0.2706	0.0306	0.2450	0.0268	0.2240	0.0239	0.2065	0.0215	0.1915
14	lib	0.0355	0.2700	0.0307	0.2444	0.0270	0.2235	0.0240	0.2060	0.0216	0.1911
15	con	0.0374	0.2650	0.0323	0.2397	0.0284	0.2191	0.0253	0.2018	0.0228	0.1872
15	lib	0.0375	0.2644	0.0325	0.2393	0.0286	0.2186	0.0254	0.2014	0.0229	0.1869
16	con	0.0392	0.2599	0.0340	0.2351	0.0299	0.2147	0.0267	0.1977	0.0240	0.1833
16	lib	0.0394	0.2593	0.0341	0.2346	0.0300	0.2143	0.0268	0.1974	0.0241	0.1829
17	con	0.0410	0.2554	0.0355	0.2308	0.0313	0.2108	0.0279	0.1940	0.0252	0.1799
17	lib	0.0412	0.2548	0.0357	0.2304	0.0314	0.2104	0.0280	0.1937	0.0253	0.1795
18	con	0.0426	0.2512	0.0370	0.2270	0.0326	0.2072	0.0291	0.1906	0.0263	0.1767
18	lib	0.0428	0.2507	0.0371	0.2266	0.0327	0.2068	0.0292	0.1903	0.0264	0.1763
19	con	0.0442	0.2475	0.0384	0.2235	0.0339	0.2039	0.0302	0.1876	0.0273	0.1738
19	lib	0.0444	0.2469	0.0385	0.2230	0.0340	0.2036	0.0304	0.1872	0.0274	0.1735
20	con	0.0456	0.2440	0.0397	0.2203	0.0350	0.2009	0.0313	0.1848	0.0283	0.1711
20	lib	0.0458	0.2434	0.0398	0.2198	0.0352	0.2005	0.0314	0.1844	0.0284	0.1708
21	con	0.0470	0.2407	0.0409	0.2173	0.0361	0.1981	0.0323	0.1822	0.0292	0.1687
21	lib	0.0472	0.2402	0.0411	0.2168	0.0363	0.1978	0.0324	0.1818	0.0293	0.1684
22	con	0.0484	0.2377	0.0421	0.2145	0.0372	0.1956	0.0333	0.1798	0.0301	0.1664
22	lib	0.0485	0.2373	0.0422	0.2141	0.0373	0.1952	0.0334	0.1795	0.0302	0.1661
23	con	0.0496	0.2350	0.0432	0.2120	0.0382	0.1932	0.0342	0.1776	0.0309	0.1643
23	lib	0.0498	0.2345	0.0433	0.2115	0.0383	0.1928	0.0343	0.1772	0.0310	0.1640
24	con	0.0508	0.2323	0.0443	0.2095	0.0391	0.1910	0.0350	0.1755	0.0317	0.1624
24	lib	0.0510	0.2319	0.0444	0.2091	0.0393	0.1906	0.0352	0.1751	0.0318	0.1621
25	con	0.0519	0.2299	0.0452	0.2073	0.0401	0.1888	0.0359	0.1735	0.0325	0.1605
25	lib	0.0521	0.2295	0.0454	0.2069	0.0402	0.1885	0.0360	0.1732	0.0326	0.1602
26	con	0.0530	0.2277	0.0462	0.2051	0.0409	0.1869	0.0367	0.1716	0.0332	0.1588
26	lib	0.0532	0.2272	0.0464	0.2048	0.0410	0.1866	0.0368	0.1714	0.0333	0.1586
27	con	0.0540	0.2255	0.0471	0.2032	0.0418	0.1850	0.0374	0.1699	0.0339	0.1572
27	lib	0.0542	0.2251	0.0473	0.2029	0.0419	0.1847	0.0375	0.1697	0.0340	0.1569
28	con	0.0550	0.2234	0.0480	0.2013	0.0425	0.1833	0.0382	0.1683	0.0345	0.1557
28	lib	0.0552	0.2231	0.0482	0.2010	0.0427	0.1830	0.0383	0.1680	0.0346	0.1554
29	con	0.0560	0.2215	0.0489	0.1995	0.0433	0.1817	0.0388	0.1668	0.0352	0.1542
29	lib	0.0561	0.2211	0.0490	0.1992	0.0434	0.1814	0.0390	0.1665	0.0353	0.1540
30	con	0.0569	0.2197	0.0497	0.1979	0.0440	0.1801	0.0395	0.1653	0.0358	0.1529
30	lib	0.0570	0.2194	0.0498	0.1975	0.0442	0.1798	0.0396	0.1650	0.0359	0.1527
31	con	0.0577	0.2180	0.0505	0.1963	0.0447	0.1787	0.0401	0.1640	0.0364	0.1516
31	lib	0.0579	0.2176	0.0506	0.1960	0.0449	0.1783	0.0403	0.1637	0.0365	0.1514
32	con	0.0586	0.2164	0.0512	0.1948	0.0454	0.1772	0.0408	0.1626	0.0369	0.1504
32	lib	0.0588	0.2160	0.0513	0.1945	0.0455	0.1770	0.0409	0.1624	0.0371	0.1501
33	con	0.0594	0.2149	0.0519	0.1934	0.0461	0.1759	0.0414	0.1614	0.0375	0.1492
33	lib	0.0595	0.2145	0.0521	0.1930	0.0462	0.1757	0.0415	0.1612	0.0376	0.1490
34	con	0.0601	0.2134	0.0526	0.1920	0.0467	0.1747	0.0419	0.1603	0.0380	0.1481
34	lib	0.0603	0.2130	0.0528	0.1917	0.0468	0.1744	0.0420	0.1600	0.0381	0.1478

Table B.4. (continued) *ANOMV Critical Values for* $\alpha = 0.01$.

						Number of Variances (k)						
	3			4		5		6		7		
ν	lower	upper		lower	upper	lower	upper	lower	upper	lower	upper	
3	0.0055	0.9064	con	0.0030	0.8130	0.0020	0.7311	0.0014	0.6601	0.0011	0.6042	
3			lib	0.0030	0.8126	0.0020	0.7305	0.0014	0.6595	0.0011	0.6042	
4	0.0134	0.8592	con	0.0079	0.7531	0.0054	0.6656	0.0040	0.5952	0.0031	0.5383	
4			lib	0.0079	0.7522	0.0054	0.6649	0.0040	0.5949	0.0031	0.5378	
5	0.0230	0.8197	con	0.0140	0.7065	0.0098	0.6176	0.0074	0.5484	0.0058	0.4926	
5			lib	0.0141	0.7062	0.0098	0.6172	0.0074	0.5478	0.0058	0.4926	
6	0.0329	0.7868	con	0.0206	0.6702	0.0146	0.5810	0.0112	0.5130	0.0089	0.4589	
6			lib	0.0207	0.6696	0.0147	0.5809	0.0112	0.5126	0.0089	0.4589	
7	0.0426	0.7592	con	0.0271	0.6405	0.0195	0.5521	0.0150	0.4852	0.0121	0.4329	
7			lib	0.0272	0.6400	0.0196	0.5518	0.0151	0.4850	0.0121	0.4327	
8	0.0518	0.7358	con	0.0334	0.6161	0.0243	0.5284	0.0189	0.4628	0.0153	0.4121	
8			lib	0.0336	0.6156	0.0244	0.5282	0.0189	0.4625	0.0153	0.4117	
9	0.0604	0.7156	con	0.0394	0.5954	0.0289	0.5088	0.0225	0.4442	0.0183	0.3945	
9			lib	0.0396	0.5949	0.0290	0.5083	0.0226	0.4441	0.0184	0.3944	
10	0.0684	0.6980	con	0.0451	0.5777	0.0332	0.4919	0.0260	0.4285	0.0212	0.3800	
10			lib	0.0452	0.5772	0.0333	0.4916	0.0261	0.4284	0.0213	0.3799	
11	0.0758	0.6824	con	0.0504	0.5623	0.0373	0.4775	0.0293	0.4150	0.0240	0.3674	
11			lib	0.0505	0.5619	0.0374	0.4771	0.0294	0.4149	0.0241	0.3673	
12	0.0827	0.6687	con	0.0553	0.5488	0.0411	0.4648	0.0324	0.4034	0.0266	0.3566	
12			lib	0.0554	0.5483	0.0412	0.4645	0.0325	0.4032	0.0267	0.3565	
13	0.0891	0.6563	con	0.0599	0.5368	0.0447	0.4537	0.0354	0.3931	0.0291	0.3471	
13			lib	0.0600	0.5364	0.0448	0.4534	0.0354	0.3928	0.0291	0.3470	
14	0.0950	0.6452	con	0.0642	0.5261	0.0481	0.4437	0.0381	0.3839	0.0314	0.3387	
14			lib	0.0644	0.5256	0.0482	0.4434	0.0382	0.3838	0.0315	0.3386	
15	0.1006	0.6350	con	0.0682	0.5164	0.0512	0.4348	0.0407	0.3758	0.0336	0.3312	
15			lib	0.0684	0.5160	0.0513	0.4345	0.0408	0.3756	0.0337	0.3311	
16	0.1058	0.6258	con	0.0721	0.5076	0.0542	0.4267	0.0432	0.3684	0.0357	0.3243	
16			lib	0.0722	0.5072	0.0543	0.4264	0.0432	0.3682	0.0358	0.3242	
17	0.1107	0.6174	con	0.0756	0.4997	0.0570	0.4194	0.0455	0.3616	0.0377	0.3183	
17			lib	0.0758	0.4992	0.0571	0.4191	0.0456	0.3615	0.0377	0.3181	
18	0.1153	0.6096	con	0.0790	0.4923	0.0597	0.4127	0.0477	0.3555	0.0396	0.3126	
18			lib	0.0792	0.4919	0.0598	0.4124	0.0478	0.3554	0.0396	0.3125	
19	0.1196	0.6024	con	0.0822	0.4856	0.0622	0.4065	0.0498	0.3500	0.0413	0.3075	
19			lib	0.0824	0.4852	0.0623	0.4063	0.0498	0.3498	0.0414	0.3074	
20	0.1236	0.5958	con	0.0852	0.4793	0.0646	0.4008	0.0518	0.3448	0.0430	0.3028	
20			lib	0.0854	0.4790	0.0647	0.4006	0.0518	0.3446	0.0431	0.3027	
21	0.1275	0.5896	con	0.0881	0.4736	0.0669	0.3956	0.0536	0.3401	0.0446	0.2984	
21			lib	0.0882	0.4732	0.0670	0.3954	0.0537	0.3398	0.0447	0.2983	
22	0.1311	0.5837	con	0.0908	0.4682	0.0691	0.3907	0.0554	0.3356	0.0462	0.2944	
22			lib	0.0910	0.4678	0.0691	0.3905	0.0555	0.3354	0.0462	0.2943	
23	0.1346	0.5783	con	0.0934	0.4631	0.0711	0.3862	0.0571	0.3315	0.0476	0.2906	
23			lib	0.0935	0.4628	0.0712	0.3859	0.0572	0.3313	0.0476	0.2905	
24	0.1379	0.5731	con	0.0958	0.4584	0.0731	0.3819	0.0588	0.3276	0.0490	0.2872	
24			lib	0.0960	0.4581	0.0731	0.3817	0.0588	0.3275	0.0490	0.2870	
25	0.1410	0.5683	con	0.0982	0.4540	0.0749	0.3779	0.0603	0.3240	0.0503	0.2839	
25			lib	0.0983	0.4537	0.0750	0.3777	0.0604	0.3239	0.0504	0.2838	
26	0.1439	0.5638	con	0.1004	0.4498	0.0767	0.3741	0.0618	0.3206	0.0516	0.2808	
26			lib	0.1006	0.4495	0.0768	0.3739	0.0619	0.3205	0.0516	0.2807	
27	0.1468	0.5595	con	0.1025	0.4459	0.0784	0.3706	0.0632	0.3175	0.0528	0.2779	
27			lib	0.1027	0.4456	0.0785	0.3704	0.0633	0.3173	0.0529	0.2778	
28	0.1495	0.5554	con	0.1046	0.4422	0.0801	0.3673	0.0646	0.3144	0.0540	0.2752	
28			lib	0.1048	0.4418	0.0802	0.3671	0.0647	0.3143	0.0540	0.2750	
29	0.1521	0.5515	con	0.1065	0.4387	0.0816	0.3642	0.0659	0.3116	0.0551	0.2725	
29			lib	0.1067	0.4383	0.0817	0.3640	0.0660	0.3115	0.0552	0.2724	
30	0.1546	0.5479	con	0.1084	0.4353	0.0832	0.3611	0.0672	0.3090	0.0562	0.2701	
30			lib	0.1086	0.4350	0.0833	0.3609	0.0672	0.3088	0.0563	0.2700	
31	0.1570	0.5444	con	0.1103	0.4321	0.0846	0.3583	0.0684	0.3064	0.0572	0.2678	
31			lib	0.1104	0.4318	0.0847	0.3581	0.0685	0.3062	0.0573	0.2677	
32	0.1593	0.5410	con	0.1120	0.4291	0.0860	0.3556	0.0696	0.3039	0.0583	0.2656	
32			lib	0.1122	0.4288	0.0861	0.3554	0.0696	0.3038	0.0583	0.2655	
33	0.1615	0.5378	con	0.1137	0.4262	0.0874	0.3530	0.0707	0.3016	0.0592	0.2635	
33			lib	0.1138	0.4259	0.0875	0.3528	0.0708	0.3015	0.0593	0.2634	
34	0.1636	0.5348	con	0.1153	0.4235	0.0887	0.3506	0.0718	0.2994	0.0602	0.2615	
34			lib	0.1154	0.4232	0.0888	0.3504	0.0718	0.2993	0.0602	0.2614	

Table B.4. (continued) *ANOMV Critical Values for $\alpha = 0.01$.*

		Number of Variances (k)								
		8		9		10		11		12	
ν		lower	upper	lower	upper	lower	upper	lower	upper	lower	upper
3	con	0.0009	0.5551	0.0007	0.5141	0.0006	0.4807	0.0005	0.4486	0.0004	0.4185
3	lib	0.0009	0.5551	0.0007	0.5135	0.0006	0.4802	0.0005	0.4482	0.0004	0.4185
4	con	0.0025	0.4914	0.0021	0.4528	0.0018	0.4193	0.0015	0.3901	0.0013	0.3650
4	lib	0.0025	0.4909	0.0021	0.4528	0.0018	0.4193	0.0015	0.3897	0.0013	0.3646
5	con	0.0048	0.4481	0.0040	0.4106	0.0034	0.3788	0.0030	0.3533	0.0026	0.3291
5	lib	0.0048	0.4480	0.0040	0.4102	0.0034	0.3788	0.0030	0.3530	0.0026	0.3290
6	con	0.0074	0.4161	0.0062	0.3801	0.0054	0.3508	0.0047	0.3255	0.0042	0.3036
6	lib	0.0074	0.4160	0.0062	0.3801	0.0054	0.3508	0.0047	0.3252	0.0042	0.3036
7	con	0.0101	0.3913	0.0086	0.3570	0.0074	0.3284	0.0065	0.3043	0.0058	0.2839
7	lib	0.0101	0.3912	0.0086	0.3570	0.0074	0.3281	0.0065	0.3040	0.0058	0.2836
8	con	0.0127	0.3715	0.0109	0.3386	0.0095	0.3110	0.0083	0.2877	0.0074	0.2679
8	lib	0.0128	0.3714	0.0109	0.3382	0.0095	0.3110	0.0083	0.2877	0.0074	0.2679
9	con	0.0154	0.3552	0.0132	0.3231	0.0115	0.2967	0.0101	0.2742	0.0091	0.2552
9	lib	0.0154	0.3551	0.0132	0.3231	0.0115	0.2966	0.0101	0.2742	0.0091	0.2552
10	con	0.0179	0.3415	0.0153	0.3104	0.0134	0.2846	0.0119	0.2631	0.0106	0.2444
10	lib	0.0179	0.3415	0.0154	0.3103	0.0134	0.2844	0.0119	0.2630	0.0106	0.2444
11	con	0.0202	0.3299	0.0174	0.2995	0.0152	0.2745	0.0135	0.2533	0.0121	0.2354
11	lib	0.0203	0.3298	0.0175	0.2994	0.0153	0.2744	0.0135	0.2533	0.0122	0.2353
12	con	0.0225	0.3199	0.0194	0.2902	0.0170	0.2656	0.0151	0.2450	0.0136	0.2277
12	lib	0.0225	0.3196	0.0194	0.2900	0.0170	0.2655	0.0151	0.2450	0.0136	0.2276
13	con	0.0246	0.3110	0.0213	0.2820	0.0187	0.2581	0.0166	0.2378	0.0149	0.2208
13	lib	0.0247	0.3109	0.0213	0.2818	0.0187	0.2579	0.0166	0.2378	0.0150	0.2207
14	con	0.0266	0.3033	0.0231	0.2747	0.0203	0.2512	0.0180	0.2315	0.0162	0.2148
14	lib	0.0267	0.3032	0.0231	0.2747	0.0203	0.2512	0.0181	0.2314	0.0163	0.2147
15	con	0.0285	0.2963	0.0247	0.2682	0.0218	0.2452	0.0194	0.2259	0.0175	0.2095
15	lib	0.0286	0.2962	0.0248	0.2681	0.0218	0.2451	0.0194	0.2257	0.0175	0.2094
16	con	0.0304	0.2900	0.0263	0.2623	0.0232	0.2398	0.0207	0.2208	0.0186	0.2047
16	lib	0.0304	0.2899	0.0263	0.2623	0.0232	0.2396	0.0207	0.2207	0.0187	0.2046
17	con	0.0321	0.2844	0.0278	0.2571	0.0245	0.2348	0.0219	0.2162	0.0198	0.2003
17	lib	0.0321	0.2842	0.0279	0.2570	0.0246	0.2347	0.0219	0.2161	0.0198	0.2003
18	con	0.0337	0.2792	0.0293	0.2524	0.0258	0.2303	0.0231	0.2120	0.0208	0.1964
18	lib	0.0337	0.2790	0.0293	0.2523	0.0258	0.2303	0.0231	0.2120	0.0208	0.1964
19	con	0.0352	0.2744	0.0306	0.2480	0.0271	0.2263	0.0242	0.2082	0.0218	0.1928
19	lib	0.0353	0.2743	0.0307	0.2479	0.0271	0.2263	0.0242	0.2082	0.0218	0.1928
20	con	0.0367	0.2702	0.0319	0.2439	0.0282	0.2226	0.0252	0.2047	0.0228	0.1896
20	lib	0.0367	0.2700	0.0319	0.2439	0.0282	0.2225	0.0252	0.2047	0.0228	0.1895
21	con	0.0381	0.2662	0.0332	0.2403	0.0293	0.2191	0.0262	0.2015	0.0237	0.1865
21	lib	0.0381	0.2660	0.0332	0.2402	0.0293	0.2191	0.0263	0.2014	0.0237	0.1865
22	con	0.0394	0.2624	0.0343	0.2369	0.0304	0.2159	0.0272	0.1985	0.0246	0.1838
22	lib	0.0395	0.2623	0.0344	0.2368	0.0304	0.2159	0.0272	0.1984	0.0246	0.1837
23	con	0.0407	0.2590	0.0355	0.2337	0.0314	0.2130	0.0281	0.1957	0.0254	0.1811
23	lib	0.0407	0.2589	0.0355	0.2336	0.0314	0.2129	0.0281	0.1956	0.0254	0.1811
24	con	0.0419	0.2557	0.0365	0.2307	0.0324	0.2102	0.0290	0.1932	0.0262	0.1787
24	lib	0.0419	0.2557	0.0366	0.2306	0.0324	0.2102	0.0290	0.1931	0.0262	0.1787
25	con	0.0431	0.2527	0.0376	0.2279	0.0333	0.2077	0.0298	0.1907	0.0270	0.1765
25	lib	0.0431	0.2526	0.0376	0.2278	0.0333	0.2076	0.0298	0.1907	0.0270	0.1764
26	con	0.0442	0.2500	0.0386	0.2253	0.0342	0.2052	0.0306	0.1885	0.0277	0.1744
26	lib	0.0442	0.2498	0.0386	0.2253	0.0342	0.2052	0.0307	0.1884	0.0278	0.1743
27	con	0.0452	0.2473	0.0395	0.2228	0.0350	0.2029	0.0314	0.1864	0.0284	0.1724
27	lib	0.0453	0.2472	0.0395	0.2228	0.0350	0.2029	0.0314	0.1863	0.0285	0.1723
28	con	0.0463	0.2448	0.0404	0.2205	0.0358	0.2008	0.0321	0.1844	0.0291	0.1705
28	lib	0.0463	0.2446	0.0404	0.2205	0.0359	0.2007	0.0322	0.1843	0.0291	0.1705
29	con	0.0473	0.2424	0.0413	0.2184	0.0366	0.1987	0.0329	0.1825	0.0298	0.1688
29	lib	0.0473	0.2423	0.0413	0.2183	0.0366	0.1987	0.0329	0.1825	0.0298	0.1687
30	con	0.0482	0.2401	0.0421	0.2163	0.0374	0.1969	0.0336	0.1807	0.0304	0.1670
30	lib	0.0482	0.2401	0.0422	0.2162	0.0374	0.1968	0.0336	0.1807	0.0304	0.1670
31	con	0.0491	0.2380	0.0429	0.2144	0.0381	0.1950	0.0342	0.1790	0.0310	0.1655
31	lib	0.0492	0.2379	0.0430	0.2143	0.0381	0.1950	0.0342	0.1790	0.0311	0.1654
32	con	0.0500	0.2360	0.0437	0.2125	0.0388	0.1933	0.0349	0.1774	0.0316	0.1639
32	lib	0.0500	0.2360	0.0438	0.2124	0.0388	0.1933	0.0349	0.1774	0.0316	0.1639
33	con	0.0509	0.2341	0.0445	0.2108	0.0395	0.1917	0.0355	0.1759	0.0322	0.1626
33	lib	0.0509	0.2341	0.0445	0.2107	0.0395	0.1916	0.0355	0.1758	0.0322	0.1625
34	con	0.0517	0.2323	0.0452	0.2091	0.0402	0.1901	0.0361	0.1745	0.0327	0.1612
34	lib	0.0517	0.2322	0.0453	0.2090	0.0402	0.1901	0.0361	0.1744	0.0328	0.1611

Table B.5. *Sample Sizes for the ANOMV* $\alpha = 0.05$

	Power= 0.80								Power= 0.90							
	I								*I*							
r	3	4	5	6	7	8	10	12	3	4	5	6	7	8	10	12
5	21	21	21	31	31	31	31	31	21	31	31	31	31	31	61	61
10	10	11	12	13	13	14	15	15	12	14	15	16	16	21	21	21
15	7	8	10	10	11	11	12	12	10	11	11	12	13	13	14	14
20	7	8	8	9	9	9	10	10	8	9	10	10	11	11	12	12
25	6	7	8	8	8	9	9	9	8	8	9	9	10	10	10	11
30	6	7	7	7	8	8	8	9	7	8	8	9	9	9	10	10
35	6	6	7	7	7	7	8	8	7	7	8	8	8	9	9	9
40	5	6	6	7	7	7	7	8	6	7	7	8	8	8	9	9

Table B.5. *Sample Sizes for the ANOMV* $\alpha = 0.1$

	Power= 0.80								Power= 0.90							
	I								*I*							
r	3	4	5	6	7	8	10	12	3	4	5	6	7	8	10	12
5	14	16	21	21	21	21	31	31	21	21	31	31	31	31	31	31
10	8	10	11	11	11	12	13	13	10	12	13	13	14	14	15	16
15	7	8	9	9	9	9	10	10	8	9	10	10	11	11	12	12
20	6	7	8	8	8	8	9	9	7	8	9	9	9	10	10	11
25	5	6	7	7	7	7	8	8	7	7	8	8	9	9	9	10
30	5	6	6	6	7	7	7	8	6	7	8	8	8	8	9	9
35	5	5	6	6	6	7	7	7	6	6	7	7	7	8	8	8
40	5	5	6	6	6	6	6	7	6	6	7	7	7	7	8	8

Table B.6. *ANOM Critical Values $g(\alpha; (I, J), \nu)$ for Two-Factor Interactions.*

	Level of Significance = 0.1					
	(I, J)					
ν	(3,3)	(3,4)	(3,5)	(4,4)	(4,5)	(5,5)
9	3.05					
10	2.99					
11	2.94					
12	2.90	3.07				
13	2.86	3.03				
14	2.83	3.00				
15	2.81	2.97	3.09			
16	2.79	2.95	3.06	3.29		
17	2.77	2.93	3.04	3.26		
18	2.75	2.91	3.02	3.24		
19	2.74	2.89	3.00	3.21		
20	2.73	2.88	2.98	3.19	3.30	
21	2.71	2.86	2.97	3.18	3.28	
22	2.70	2.85	2.96	3.16	3.26	
23	2.69	2.84	2.94	3.15	3.25	
24	2.69	2.83	2.93	3.13	3.23	
25	2.68	2.82	2.92	3.12	3.22	2.43
30	2.65	2.79	2.88	3.07	3.17	3.37
40	2.61	2.74	2.83	3.02	3.10	3.29
60	2.57	2.70	2.79	2.96	3.04	3.22
90	2.55	2.67	2.76	2.92	3.00	3.17
120	2.53	2.66	2.74	2.90	2.98	3.15
∞	2.50	2.62	2.70	2.85	2.93	3.08

Table B.6. (continued) *ANOM Critical Values $g(\alpha; (I, J), \nu)$*
for Two-Factor Interactions.

	Level of Significance = 0.05					
	\multicolumn{6}{c}{(I, J)}					
ν	(3,3)	(3,4)	(3,5)	(4,4)	(4,5)	(5,5)
9	3.50					
10	3.41					
11	3.34					
12	3.28	3.46				
13	3.24	3.41				
14	3.20	3.36				
15	3.16	3.32	3.44			
16	3.13	3.29	3.40	3.63		
17	3.11	3.26	3.37	3.59		
18	3.09	3.24	3.34	3.56		
19	3.07	3.21	3.32	3.53		
20	3.05	3.19	3.30	3.50	3.61	
21	3.03	3.18	3.28	3.48	3.58	
22	3.02	3.16	3.26	3.46	3.56	
23	3.01	3.15	3.25	3.44	3.54	
24	3.00	3.13	3.23	3.42	3.52	
25	2.98	3.12	3.22	3.41	3.50	3.71
30	2.94	3.07	3.17	3.35	3.44	3.63
40	2.89	3.02	3.10	3.28	3.36	3.54
60	2.84	2.96	3.04	3.20	3.28	3.45
90	2.81	2.92	3.00	3.16	3.23	3.40
120	2.79	2.90	2.98	3.14	3.21	3.37
∞	2.75	2.85	2.93	3.07	3.14	3.29

Table B.6. (continued) *ANOM Critical Values $g(\alpha; (I, J), \nu)$*
for Two-Factor Interactions.

	Level of Significance = 0.01					
	(I, J)					
ν	(3,3)	(3,4)	(3,5)	(4,4)	(4,5)	(5,5)
9	4.60					
10	4.43					
11	4.29					
12	4.19	4.37				
13	4.10	4.27				
14	4.03	4.19				
15	3.97	4.13	4.24			
16	3.92	4.07	4.18	4.41		
17	3.87	4.02	4.13	4.35		
18	3.83	3.98	4.08	4.29		
19	3.80	3.94	4.04	4.25		
20	3.77	3.90	4.00	4.20	4.30	
21	3.74	3.87	3.97	4.17	4.27	
22	3.71	3.85	3.94	4.13	4.23	
23	3.69	3.82	3.92	4.10	4.20	
24	3.67	3.80	3.89	4.08	4.17	
25	3.65	3.78	3.87	4.05	4.14	4.35
30	3.58	3.70	3.78	3.96	4.04	4.23
40	3.49	3.60	3.68	3.84	3.92	4.09
60	3.41	3.51	3.58	3.73	3.80	3.96
90	3.35	3.45	3.52	3.66	3.73	3.88
120	3.33	3.42	3.49	3.63	3.69	3.84
∞	3.25	3.34	3.40	3.53	3.59	3.72

Table B.7. *HANOM Critical Values* $\mathcal{H}(\alpha; k, v)$.

Level of Significance = 0.1

v	\multicolumn{19}{c}{Number of Means Being Compared, k}
	2	3	4	5	6	7	8	9	10	11	12	13	14	15	16	17	18	19	20
1	6.32	12.5	18.5	24.5	30.6	36.7	42.9	48.9	54.9	61.0	66.9	73.0	79.0	85.1	91.0	97.2	103.	109.	115.
2	2.28	3.69	4.69	5.56	6.32	7.01	7.65	8.23	8.78	9.30	9.79	10.3	10.7	11.1	11.6	11.9	12.3	12.7	13.1
3	1.75	2.68	3.24	3.70	4.08	4.42	4.71	4.98	5.23	5.46	5.67	5.87	6.05	6.22	6.39	6.55	6.70	6.85	6.99
4	1.55	2.33	2.76	3.10	3.37	3.61	3.81	3.99	4.16	4.30	4.44	4.57	4.69	4.80	4.91	5.00	5.10	5.19	5.27
5	1.45	2.16	2.53	2.81	3.04	3.23	3.39	3.53	3.66	3.78	3.88	3.98	4.07	4.16	4.24	4.32	4.39	4.45	4.52
6	1.39	2.06	2.40	2.65	2.85	3.01	3.15	3.27	3.38	3.48	3.57	3.65	3.73	3.80	3.87	3.93	3.99	4.05	4.10
7	1.36	1.99	2.31	2.54	2.72	2.87	3.00	3.11	3.20	3.29	3.37	3.45	3.52	3.58	3.64	3.69	3.74	3.79	3.84
8	1.33	1.94	2.25	2.46	2.63	2.77	2.89	2.99	3.08	3.16	3.23	3.30	3.36	3.42	3.47	3.52	3.57	3.61	3.66
9	1.31	1.91	2.20	2.41	2.57	2.70	2.81	2.91	2.99	3.06	3.13	3.20	3.25	3.31	3.35	3.40	3.44	3.48	3.52
10	1.29	1.88	2.17	2.37	2.52	2.65	2.75	2.84	2.92	2.99	3.06	3.12	3.17	3.22	3.26	3.31	3.35	3.39	3.42
11	1.28	1.86	2.14	2.33	2.48	2.60	2.70	2.79	2.87	2.94	3.00	3.05	3.10	3.15	3.19	3.24	3.27	3.31	3.34
12	1.27	1.84	2.12	2.31	2.45	2.57	2.67	2.75	2.82	2.89	2.95	3.00	3.05	3.10	3.14	3.18	3.21	3.25	3.28
13	1.26	1.83	2.10	2.28	2.42	2.54	2.63	2.72	2.79	2.85	2.91	2.96	3.01	3.05	3.09	3.13	3.16	3.20	3.23
14	1.25	1.82	2.08	2.26	2.40	2.51	2.61	2.69	2.76	2.82	2.88	2.93	2.97	3.01	3.05	3.09	3.12	3.16	3.19
15	1.24	1.81	2.07	2.25	2.38	2.49	2.59	2.66	2.73	2.79	2.85	2.90	2.94	2.98	3.02	3.06	3.09	3.12	3.15
16	1.24	1.80	2.06	2.23	2.37	2.48	2.57	2.65	2.71	2.77	2.82	2.87	2.91	2.96	2.99	3.03	3.06	3.09	3.12
17	1.23	1.79	2.05	2.22	2.35	2.46	2.55	2.63	2.69	2.75	2.80	2.85	2.89	2.93	2.97	3.00	3.03	3.07	3.09
18	1.23	1.78	2.04	2.21	2.34	2.45	2.54	2.61	2.68	2.73	2.79	2.83	2.87	2.91	2.95	2.98	3.01	3.04	3.07
19	1.23	1.78	2.03	2.20	2.33	2.44	2.52	2.60	2.66	2.72	2.77	2.81	2.85	2.89	2.93	2.96	2.99	3.02	3.05
20	1.22	1.77	2.02	2.19	2.32	2.43	2.51	2.59	2.65	2.70	2.75	2.80	2.84	2.88	2.91	2.95	2.98	3.00	3.03
24	1.21	1.76	2.00	2.17	2.29	2.39	2.48	2.55	2.61	2.66	2.71	2.75	2.79	2.83	2.86	2.89	2.92	2.95	2.97
30	1.20	1.74	1.98	2.14	2.26	2.36	2.44	2.51	2.57	2.62	2.67	2.71	2.75	2.78	2.81	2.84	2.87	2.89	2.92
40	1.19	1.72	1.96	2.12	2.24	2.33	2.41	2.47	2.53	2.58	2.62	2.66	2.70	2.73	2.76	2.79	2.82	2.84	2.87
60	1.18	1.71	1.94	2.09	2.21	2.30	2.38	2.44	2.49	2.54	2.58	2.62	2.66	2.69	2.72	2.74	2.77	2.79	2.81
120	1.17	1.69	1.92	2.07	2.18	2.27	2.34	2.40	2.46	2.50	2.54	2.58	2.61	2.64	2.67	2.70	2.72	2.74	2.77
∞	1.16	1.68	1.90	2.05	2.15	2.24	2.31	2.37	2.42	2.47	2.50	2.54	2.57	2.60	2.63	2.65	2.68	2.70	2.72

Table B.7. (continued) *HANOM Critical Values* $\mathcal{H}(\alpha; k, \nu)$.

Level of Significance = 0.05

ν	\multicolumn{19}{c}{Number of Means Being Compared, k}
	2	3	4	5	6	7	8	9	10	11	12	13	14	15	16	17	18	19	20
1	12.7	25.3	37.8	50.1	62.5	74.7	87.1	99.4	112.	125.	137.	150.	162.	175.	187.	200.	212.	225.	237.
2	3.28	5.24	6.72	7.97	9.07	10.0	11.0	11.8	12.6	13.3	14.0	14.7	15.35	16.0	16.6	17.2	17.7	18.3	18.8
3	2.29	3.43	4.17	4.75	5.25	5.67	6.05	6.40	6.71	7.00	7.28	7.53	7.76	7.99	8.20	8.39	8.58	8.77	8.94
4	1.97	2.86	3.39	3.80	4.13	4.41	4.65	4.87	5.07	5.25	5.41	5.57	5.70	5.83	5.96	6.08	6.19	6.30	6.40
5	1.81	2.60	3.04	3.37	3.62	3.84	4.03	4.20	4.35	4.49	4.61	4.72	4.82	4.91	5.01	5.09	5.18	5.25	5.33
6	1.72	2.45	2.83	3.12	3.34	3.52	3.68	3.82	3.95	4.06	4.16	4.25	4.34	4.42	4.49	4.56	4.62	4.69	4.75
7	1.66	2.35	2.71	2.96	3.16	3.33	3.46	3.59	3.70	3.80	3.89	3.97	4.04	4.11	4.17	4.23	4.29	4.34	4.39
8	1.62	2.29	2.62	2.86	3.04	3.19	3.32	3.43	3.53	3.62	3.69	3.77	3.83	3.89	3.95	4.00	4.06	4.10	4.15
9	1.59	2.24	2.55	2.78	2.95	3.09	3.21	3.31	3.40	3.48	3.56	3.62	3.69	3.74	3.79	3.84	3.89	3.93	3.97
10	1.56	2.20	2.50	2.71	2.88	3.01	3.13	3.22	3.31	3.39	3.45	3.52	3.57	3.63	3.67	3.72	3.76	3.80	3.84
11	1.55	2.17	2.46	2.67	2.83	2.96	3.06	3.15	3.24	3.31	3.37	3.43	3.49	3.54	3.58	3.62	3.66	3.70	3.74
12	1.53	2.14	2.43	2.63	2.78	2.91	3.01	3.10	3.18	3.24	3.31	3.36	3.41	3.46	3.51	3.55	3.58	3.62	3.65
13	1.52	2.12	2.40	2.60	2.75	2.87	2.97	3.06	3.13	3.19	3.25	3.31	3.36	3.40	3.44	3.48	3.52	3.56	3.59
14	1.51	2.10	2.38	2.57	2.72	2.83	2.93	3.02	3.09	3.15	3.21	3.26	3.31	3.35	3.39	3.43	3.47	3.50	3.53
15	1.50	2.09	2.36	2.55	2.69	2.81	2.90	2.98	3.05	3.12	3.17	3.22	3.27	3.31	3.35	3.39	3.42	3.45	3.48
16	1.49	2.08	2.35	2.53	2.67	2.78	2.88	2.96	3.02	3.09	3.14	3.19	3.24	3.28	3.32	3.35	3.38	3.42	3.44
17	1.48	2.07	2.33	2.52	2.65	2.76	2.85	2.93	3.00	3.06	3.11	3.16	3.21	3.25	3.28	3.32	3.35	3.38	3.41
18	1.48	2.06	2.32	2.50	2.64	2.75	2.84	2.91	2.98	3.04	3.09	3.14	3.18	3.22	3.26	3.29	3.32	3.35	3.38
19	1.47	2.05	2.31	2.49	2.62	2.73	2.82	2.89	2.96	3.02	3.07	3.12	3.16	3.20	3.23	3.27	3.30	3.33	3.36
20	1.47	2.04	2.30	2.48	2.61	2.72	2.80	2.88	2.94	3.00	3.05	3.10	3.14	3.17	3.21	3.25	3.28	3.31	3.33
24	1.45	2.02	2.27	2.44	2.57	2.67	2.76	2.83	2.89	2.94	2.99	3.04	3.08	3.11	3.15	3.18	3.21	3.23	3.26
30	1.44	2.00	2.24	2.41	2.53	2.63	2.71	2.78	2.84	2.89	2.94	2.98	3.01	3.05	3.08	3.11	3.14	3.16	3.19
40	1.43	1.98	2.22	2.38	2.50	2.59	2.67	2.73	2.79	2.84	2.88	2.92	2.96	2.99	3.02	3.05	3.07	3.10	3.12
60	1.41	1.95	2.19	2.35	2.46	2.55	2.63	2.69	2.74	2.79	2.83	2.87	2.90	2.93	2.96	2.99	3.01	3.04	3.06
120	1.40	1.93	2.16	2.31	2.43	2.51	2.58	2.64	2.70	2.74	2.78	2.82	2.85	2.88	2.90	2.93	2.95	2.97	3.00
∞	1.39	1.91	2.14	2.29	2.39	2.48	2.54	2.60	2.65	2.70	2.73	2.77	2.80	2.83	2.85	2.88	2.90	2.92	2.94

Table B.7. (continued) *HANOM Critical Values* $\mathcal{H}(\alpha; k, \nu)$.

Level of Significance = 0.01

Number of Means Being Compared, k

ν	2	3	4	5	6	7	8	9	10	11	12	13	14	15	16	17	18	19	20
1	63.6	128.	190.	253.	318.	381.	446.	508.	570.	634.	698.	758.	824.	885.	950.	1015.	1077.	1145.	1210.
2	7.24	11.7	15.1	18.0	20.5	22.7	24.9	26.8	28.5	30.2	31.9	33.3	34.8	36.3	37.7	38.9	40.2	41.4	42.6
3	3.95	5.89	7.21	8.24	9.15	9.88	10.5	11.1	11.7	12.2	12.6	13.1	13.4	13.8	14.2	14.5	14.8	15.2	15.5
4	3.07	4.37	5.20	5.83	6.34	6.77	7.15	7.49	7.78	8.05	8.30	8.53	8.75	8.95	9.13	9.31	9.47	9.64	9.80
5	2.69	3.74	4.38	4.85	5.22	5.53	5.79	6.04	6.24	6.43	6.59	6.75	6.89	7.03	7.15	7.27	7.39	7.50	7.59
6	2.48	3.41	3.93	4.32	4.63	4.87	5.08	5.26	5.42	5.58	5.71	5.83	5.94	6.04	6.13	6.23	6.31	6.39	6.46
7	2.35	3.20	3.66	4.01	4.27	4.47	4.65	4.81	4.95	5.08	5.19	5.28	5.38	5.46	5.54	5.61	5.69	5.76	5.82
8	2.27	3.06	3.48	3.79	4.02	4.21	4.36	4.50	4.63	4.73	4.83	4.92	4.99	5.07	5.14	5.20	5.26	5.32	5.38
9	2.20	2.96	3.35	3.63	3.84	4.01	4.16	4.29	4.40	4.50	4.59	4.66	4.74	4.80	4.86	4.92	4.97	5.02	5.08
10	2.15	2.88	3.25	3.52	3.72	3.87	4.01	4.13	4.23	4.32	4.40	4.47	4.53	4.59	4.65	4.70	4.75	4.79	4.85
11	2.12	2.83	3.18	3.42	3.61	3.77	3.89	4.00	4.10	4.18	4.25	4.33	4.38	4.44	4.49	4.54	4.59	4.63	4.66
12	2.08	2.78	3.12	3.36	3.53	3.67	3.80	3.90	3.99	4.07	4.14	4.20	4.26	4.31	4.36	4.41	4.45	4.49	4.53
13	2.06	2.74	3.07	3.30	3.47	3.61	3.72	3.82	3.91	3.98	4.05	4.11	4.17	4.22	4.26	4.30	4.35	4.38	4.42
14	2.04	2.71	3.03	3.25	3.42	3.55	3.66	3.75	3.83	3.91	3.97	4.03	4.08	4.13	4.17	4.21	4.25	4.29	4.33
15	2.02	2.69	3.00	3.21	3.37	3.50	3.61	3.70	3.78	3.85	3.91	3.96	4.01	4.06	4.11	4.14	4.18	4.22	4.25
16	2.01	2.67	2.97	3.17	3.34	3.46	3.56	3.65	3.72	3.79	3.85	3.91	3.95	4.00	4.04	4.08	4.11	4.15	4.18
17	1.99	2.64	2.94	3.14	3.30	3.42	3.52	3.61	3.68	3.75	3.81	3.86	3.91	3.95	3.99	4.02	4.06	4.09	4.13
18	1.98	2.62	2.93	3.12	3.27	3.39	3.49	3.58	3.65	3.71	3.77	3.82	3.87	3.91	3.94	3.98	4.01	4.05	4.08
19	1.98	2.61	2.90	3.10	3.25	3.37	3.46	3.54	3.62	3.68	3.73	3.78	3.83	3.87	3.91	3.94	3.97	4.01	4.04
20	1.96	2.59	2.89	3.08	3.23	3.34	3.44	3.52	3.59	3.65	3.71	3.75	3.80	3.83	3.87	3.91	3.94	3.97	4.00
24	1.94	2.55	2.83	3.02	3.16	3.26	3.36	3.43	3.50	3.56	3.61	3.65	3.70	3.73	3.77	3.80	3.83	3.86	3.89
30	1.91	2.51	2.79	2.96	3.09	3.20	3.28	3.35	3.41	3.47	3.52	3.56	3.60	3.63	3.67	3.70	3.73	3.75	3.77
40	1.89	2.48	2.74	2.91	3.03	3.13	3.21	3.27	3.33	3.38	3.43	3.47	3.51	3.54	3.57	3.60	3.62	3.65	3.67
60	1.86	2.44	2.69	2.85	2.97	3.07	3.14	3.20	3.26	3.31	3.35	3.38	3.42	3.45	3.47	3.50	3.53	3.55	3.57
120	1.84	2.41	2.64	2.80	2.92	3.00	3.07	3.13	3.18	3.23	3.27	3.30	3.33	3.36	3.39	3.41	3.44	3.45	3.48
∞	1.82	2.38	2.61	2.76	2.87	2.95	3.01	3.07	3.12	3.16	3.20	3.23	3.26	3.28	3.31	3.33	3.35	3.37	3.39

Appendix C
SAS Examples

C.1   Introduction

This appendix contains several SAS programs that illustrate the SAS system's ANOM capabilities and how SAS was used to produce selected ANOM decision charts appearing in this book. More extensive descriptions can be found at the SAS website http://www.sas.com and in the *SAS/QC User's Guide Version* 9 (2003) published by SAS Institute, Inc.

C.2   Examples from Chapter 2

The SAS program in Table C.1 produces the normal probability plot in Figure 2.3 and the ANOM decision chart in Figure 2.4. PROC GLM is used to create a file containing the residuals (resplotc2fweight), and PROC CAPABILITY is used to produce the probability plot. PROC ANOM is then used to create the ANOM decision chart. Note that "xchart" is used since we wish to use the normal data ANOM procedure. The "alpha = .01" option specifies the level of significance. Unlike lookup methods, using SAS allows users to choose any level of significance (as opposed to tables which have only a small subset of alphas).

The SAS program in Table C.2 illustrates using SAS to determine the p-value for an ANOM decision chart. A trial and error method was used until $\alpha = 0.0099$ produced the appropriate chart. (A decision line intersects the most extreme mean.) The SAS program produces the ANOM chart in Figure 2.10.

The SAS program in Table C.3 creates a data file for the office visit data from Example 2.2 and produces the ANOM chart in Figure 2.11. The variable Clinic identifies the clinic, and the variable NumOnTime identifies the number on time. Since this example is for binomial data, the pchart command for PROC ANOM is used. The statement "total = 60" in the data step identifies the sample size for each subgroup and is identified in PROC ANOM by the "groupn = total" statement.

The SAS program in Table C.4 creates a data file for the arrival rate data from Example 2.15 and produces the ANOM chart in Figure 2.13. The variable Clinic identifies the clinic, and the variable Arrivals identifies the number of arrivals per unit of time. Since this example is for Poisson data, the uchart command for PROC ANOM is used. The statement "total =

60" in the data step identifies the sample size (area of opportunity) for each subgroup and is connected in PROC ANOM by the "groupn = total" statement.

C.3   Examples from Chapter 3

In SAS, unbalanced ANOM is handled in almost the same manner as balanced ANOM. The SAS program in Table C.5 produces the ANOM chart in Figure 3.5. Since there are three samples from location A and five from location B, the design is unbalanced. SAS automatically takes this into account when determining the decision lines from raw data. If the data is summarized then details about subgroup sample sizes must be provided.

The SAS program in Table C.6 produces the ANOM chart in Figure 3.7. Since binomial data is involved, the pchart option is used. The variable Enrollment specifies the sample size for each school and is identified in the "groupn = Enrollment" statement in PROC ANOM. This example illustrates the use of the OUTTABLE option. The statement "outtable = MathSummary" creates an output file containing the information in Table C.7, which can be used to identify the decision limits with greater accuracy.

The SAS program in Table C.8 creates a data file and the ANOM chart in Figure 3.8. The variable resvac identifies the resin and whether the vacuum is off or on. Since Poisson data are involved, the uchart option is used. The variable Npanels specifies the sample size (area of opportunity) for each subgroup and is identified in the "groupn = Npanels" statement in PROC ANOM.

C.4   Examples from Chapter 5

The SAS program in Table C.9 produces the ANOVA output in Table 5.5 for the hemoglobin data in Example 5.2. Since there is no evidence of interaction, main effects analyses can be performed. The ANOVA table can be used to get MS_e and the df for the ANOM decision charts, which are produced by the two PROC ANOM statements. Note that in both of these, $MS_e = 2.110$ and df $= 24$ are specified in the "xchart" statement. The first PROC ANOM produces Figure 5.12.

When interaction is significant, there are two approaches. One may do a simple effects analysis in which an ANOM chart is produced for each level of one of the two factors. In this case, one decision chart is produced for the UV case and another decision chart for the No-UV case. This can be accomplished using a "by UV" statement, which is illustrated in Table C.11 with the PROC ANOM statement. The use of by UV produces a decision chart for each level of the variable UV. This SAS code produces the chart in Figure 5.14 as well as the chart for the No-UV case. The outtable option produces the file pyldsum1, which contains precise decision limits and other information. Using the statement "by Monomer" would produce three decision charts, one for each level of the factor monomer. (Note that the xchart statement would then be "xchart y*UV" and the data would be sorted by monomer instead of UV.)

Alternatively, when interaction is present, one can treat the data as if it were a one-way layout by associating each level of the combined factor with the combinations of the factor levels. The SAS variable MonUV created in the data step (see Table C.10) achieves this purpose. The xchart y*MonUV in the PROC ANOM statement in Table C.12 produces the decision chart in Figure 5.15.

Table C.1. *SAS Program for Fill Weight Data (Example* 2.1*).*

```
data c2fweight; input Treatment$  Weight;
 datalines;
 1    119.5
 1    119.9
 1    120.4
 1    121.2
 1    118.7
 1    119.3
 1    119.6
 2    122.1
 2    123.9
 2    123.6
 2    122.4
 2    122.7
 2    123.1
 2    122.8
 3    120.4
 3    120.8
 3    121.4
 3    122.8
 3    122.5
 3    121.9
 3    123.1
 4    120.7
 4    122.7
 4    119.7
 4    120.7
 4    121.4
 4    121.6
 4    121.3
 ;

title 'Fill Weight GLM for Residuals';
proc glm data=c2fweight;
     class treatment;
     model weight = treatment;
     output out = resplotc2fweight p=pred r=residual;
run;

title 'Normal Probability Plot Fill Weight Data Residuals';

proc capability data = resplotc2fweight;
     qqplot residual / normal;
run;
proc anom data=C2fweight;
     xchart weight*treatment/ alpha=.01  ;
     title 'Treatment Effect ANOM Chart for Fill Weight Data';
     label weight = 'Mean Weight'
           treatment = 'Treatments';
run;
```

Table C.2. *SAS Program for Injection Depth Data (Example 2.9).*

```
data C2injection;
input Location$   Depth      Depth_centered;
 datalines;
 buttocks     17.3     0.025
 buttocks     18.5     1.225
 buttocks     17.5     0.225
 buttocks     15.8    -1.475
 post_thigh   18.7     1.65
 post_thigh   15.8    -1.25
 post_thigh   16.5    -0.55
 post_thigh   17.2     0.15
 thigh        22.6     2.075
 thigh        19.6    -0.925
 thigh        21.4     0.875
 thigh        18.5    -2.025 ;

proc anom data=C2injection;
     xchart depth*location/ alpha=.0099   ;
     title 'ANOM Chart Illustrating P-Value (Injection Depth Data)';
     label depth = 'Mean Depth'
           location = 'Location';
   run;
```

Table C.3. *SAS Program for On-Time Office Visit Data (Example 2.2).*

```
data c2offvisits; input Clinic$ NumOnTime;
 total = 60;
 datalines;
 A    10
 B    48
 C    37
 D    8
 E    35
 F    50 ;

proc anom data=c2offvisits;
     pchart NumOnTime*Clinic/ alpha=.01 groupn = total ;
     title 'Clinic Effect ANOM Chart for Proportion On Time';
     label NumOnTime = 'Proportion on Time'
           Clinic = 'Clinic';
   run;
```

Table C.4. *SAS Program for Urgent Care Arrival Rates Data (Example 2.15).*

```
data c2Carrivals; input Clinic$ Arrivals;
 total = 60;
 datalines;
 A    373
 B    347
 C    465
 D    388
 E    429
 F    498 ;

proc anom data=c2Carrivals;

     uchart arrivals*clinic/ alpha=.05  groupn = total;
     title 'Clinic Effect ANOM Chart for Arrival Rates';
     label arrivals = 'Arrival Rate / 60 Hours'
          Clinic = 'Clinic';
   run;
```

Table C.5. *SAS Program for Stream Remediation Data (Example 3.2).*

```
 data C3Fish; input Location$    ToxinLevel;
datalines;
A    3.8
A    3.2
A    3.5
B    5.3
B    5.7
B    5.6
B    5.8
B    5.6
C    3.3
C    2.9
C    3.8
C    3.3
C    3.4
;


    title 'Location Effect ANOM Chart for Fish Toxin Levels';
    proc anom data=C3Fish;
       xchart toxinlevel*location / alpha=.001 method=smm;
     label toxinlevel = 'Mean Toxin Level'
             location= 'Location';
    run;
```

Table C.6. *SAS Program for Standardized Test Score Data (Example* 3.4*).*

```
data C3School1;
input School$ Enrollment  N_Proficient;
 datalines;
 N1   52   50
 N2   55   15
 N3   105  89
 N4   81   75
 N5   46   40
 N6   84   67
 Alt1      18   13
 Alt2 33   28
 Alt3      48   17
 Alt4      41   21
 ;

    title 'Proportion Proficient in Math';

    proc anom data=C3School1;
       pchart N_Proficient*School / groupn = Enrollment alpha=.01
         method=smm wneedles = 0.25 outtable = MathSummary;
     label N_Proficient = 'Proportion Proficient'
             School = 'School';
    run;
proc print data = MathSummary; run;
```

Table C.7. *OUTTABLE Contents for Standardized Test Score Data (Example 3.4).*

Obs	_VAR_	School	_ALPHA_	_LIMITN_	_SUBN_	_LDLP_	_SUBP_	_UDLP_	_P_	_EXLIM_
1	N_Proficient	N1	0.01	52	52	0.54583	0.96154	0.73712	0.92842	UPPER
2	N_Proficient	N2	0.01	55	55	0.55167	0.27273	0.73712	0.92258	LOWER
3	N_Proficient	N3	0.01	105	105	0.60968	0.84762	0.73712	0.86457	
4	N_Proficient	N4	0.01	81	81	0.58826	0.92593	0.73712	0.88598	UPPER
5	N_Proficient	N5	0.01	46	46	0.53255	0.86957	0.73712	0.94170	
6	N_Proficient	N6	0.01	84	84	0.59140	0.79762	0.73712	0.88284	
7	N_Proficient	Alt1	0.01	18	18	0.40134	0.72222	0.73712	1.00000	
8	N_Proficient	Alt2	0.01	33	33	0.49257	0.84848	0.73712	0.98167	
9	N_Proficient	Alt3	0.01	48	48	0.53724	0.35417	0.73712	0.93700	LOWER
10	N_Proficient	Alt4	0.01	41	41	0.51939	0.51220	0.73712	0.95486	LOWER

Table C.8. *SAS Program for Fiberglass Data (Example 3.5).*

```
data C3Fiberglass; input  resin$ vacuum$ defects  Npanels;
 if resin = '1' and vacuum = 'Off' then resvac ='R1Voff';
 if resin = '1' and vacuum = 'On' then resvac ='R1Von';
 if resin = '2' and vacuum = 'Off' then resvac ='R2Voff';
 if resin = '2' and vacuum = 'On' then resvac ='R2Von';

datalines;

1 Off 8 32
1 On 6 32
2 Off 11 29
2 On 10 32 ;

   title 'Defect Rate by Resin/Vacuum';

   proc anom data=C3Fiberglass;
      uchart defects*resvac / groupn = Npanels alpha=.10
            method=smm wneedles = 0.25;
            label defects  = 'Defect Rate'
                  resvac = 'Resin Vacuum Combinations';
   run;
```

Table C.9. *SAS Program for Hemoglobin Data in Example 5.2.*

```
data hemo1;
    input Y A$  B$ @@;
    C = 'male';
    datalines;
11.7    1   1 13.6    1   1 14.5    1   1 16.8    1   1 11.7 1    1
15.4    1   2 14.0    1   2 12.9    1   2 15.5    1   2 12.6 1    2
16.5    1   3 15.9    1   3 15.4    1   3 17.5    1   3 16.1 1    3


9.7     2   1 12.3    2   1 12.3    2   1 11.7    2   1 13.1 2    1
13.3    2   2 11.8    2   2 11.8    2   2 12.9    2   2 9.1  2    2
13.8    2   3 12.7    2   3 12.9    2   3 11.6    2   3 14.6 2    3
;
title 'Hemoglobin Example';
   proc glm data=hemo1;
      class A B;
      model y = A·B A*B;
      output out = resplot5_1 p=pred r=residual;
      means A;
      means B;
      means A*B;
   run;
proc sort data = hemo1; by A;run;
proc anom data=hemo1;
      xchart y*A / alpha=.05 mse = 2.110 dfe = 24 wneedles=2
      vaxis = 11 12 13 14 15 16 ;
      title 'Theraphy Effect ANOM Chart for Hemoglobin Data';
      label y = 'Mean Hemoglobin'
            A = 'Levels of Factor A (Theraphy)';
run;
proc sort data = hemo1; by B;run;
proc anom data=hemo1;
      xchart y*B / alpha=.05 mse = 2.110 dfe = 24
      vaxis = 11 12 13 14 15 16 wneedles=2 ;
      title 'Drug Effect ANOM Chart for Hemoglobin Data';
      label y = 'Mean Hemoglobin'
            B = 'Levels of Factor B (Drug)';
   run;
```

Table C.10. *SAS Program Data Step for the Process Yield Data in Example* 5.3.

```
data ex5_2;
    input Y A$   B$ @@;
    if A = '1' then UV = 'UV     ';
    if A = '2' then UV = 'No UV ';
    if B = '1' then Monomer = 'M1';
    if B = '2' then Monomer = 'M2';
    if B = '3' then Monomer = 'M3';
    if A = '1' and B = '1' then MonUV = 'M1UV    ';
    if A = '1' and B = '2' then MonUV = 'M2UV    ';
    if A = '1' and B = '3' then MonUV = 'M3UV    ';
    if A = '2' and B = '1' then MonUV = 'M1NoUV ';
    if A = '2' and B = '2' then MonUV = 'M2NoUV ';
    if A = '2' and B = '3' then MonUV = 'M3NoUV ';

    datalines;
88.4    1   1 89.1    1   1 90.9    1   1 90.8    1   1 93.0   1 2
93.7    1   2 93.4    1   2 92.9    1   2 94.2    1   3 95.1   1 3
96.4    1   3 95.7    1   3 90.9    2   1 92.3    2   1 92.3   2 1
92.3    2   1 88.3    2   2 89.5    2   2 88.0    2   2 89.6   2 2
91.0    2   3 90.0    2   3 92.4    2   3 92.8    2   3 ;
```

Table C.11. *SAS Program Steps for Process Yield Data ANOM Charts by UV in Example 5.3.*

```
proc sort data = ex5_2; by UV monomer;run;

proc anom data=ex5_2;by UV;
     xchart y*monomer / alpha=0.05   wneedles=2 outtable = pyldsum1;
     title 'ANOM Chart for Process Yield Data';
     label y = 'Mean Yield'
           Monomer = 'Monomer';
  run;
proc print data = pyldsum1; run;
```

Table C.12. *SAS Program Steps for Process Yield Data ANOM Chart with Combined Factor Levels in Example 5.3.*

```
proc anom data=ex5_2;
     xchart y*MonUV / alpha=.05   wneedles=2 outtable = pyldsum2;
     title 'ANOM Chart for Process Yield Data';
     label y = 'Mean Yield'
           MonUV = 'Monomer UV Combinations';
  run;
proc print data = pyldsum2; run;
  quit;
```

References

Bakir, S. T. (1989). "Analysis of Means Using Ranks." *Communications in Statistics: Simulation and Computation* 18, pp. 757–775.

Bernard, A. J., and Wludyka, P. S. (2001). "Robust I-Sample Analysis of Means Type Randomization Tests for Variances." *Journal of Statistical Computation and Simulation* 69, pp. 57–88.

Bishop, T. A., and Dudewicz, E. J. (1978). "Exact Analysis of Variance with Unequal Variances: Test Procedures and Tables." *Technometrics* 20, pp. 419–430.

Bishop, T. A., and Dudewicz, E. J. (1981). "Heteroscedastic ANOVA." *Sankhyā, Series B* 43, pp. 40–57.

Box, G. E. P., Hunter, W. G., and Hunter, J. S. (1978). *Statistics for Experimenters.* John Wiley & Sons, New York.

Conover, W. J., Johnson, M. E., and Johnson, M. M. (1981). "A Comparative Study of Tests for Homogeneity of Variances with Application to the Outer Continental Shelf Bidding Data." *Technometrics* 23, pp. 351–361.

Cornell, J. A. (2002). *Experiments with Mixtures: Designs, Models, and the Analysis of Mixture Data*, 3rd ed. John Wiley & Sons, New York.

Cornell, J. A. (1983). *How to Run Mixture Experiments for Product Quality.* American Society for Quality, Milwaukee, WI.

Craig, C. D. (1947). "Control Charts Versus the Analysis of Variance in Process Control by Variables." *Industrial Quality Control* 3, pp. 14–16.

Davies, O. L., ed. (1978). *The Design and Analysis of Industrial Experiments.* Longman Inc., New York.

Halperin, M., Greenhouse, S. W., Cornfield, J., and Zalokar, J. (1955). "Tables of Percentage Points for the Studentized Maximum Absolute Deviate in Normal Samples." *Journal of the American Statistical Association* 50, pp. 185–195.

Dudewicz, E. J., and Mishra, S. N. (1988). *Modern Mathematical Statistics.* John Wiley & Sons, New York.

Dudewicz, E. J., and Nelson, P. R. (2003). "Heteroscedastic Analysis of Means (HANOM)." *American Journal of Mathematical and Management Sciences* 23, pp. 143–181.

Enrick, N. L. (1976). "An Analysis of Means of a Three-Way Factorial." *Journal of Quality Technology* 8, pp. 189–196.

Enrick, N. L. (1981). "Analysis of Means in Market Research." *Journal of the Academy of Marketing Science* 9(4), pp. 368–379.

Farnum, N. R. (1994). *Modern Statistical Quality Control and Improvement*. Duxbury Press, Belmont, CA.

Fisher, R. A. (1918). "The Correlation Between Relatives on the Supposition of Mendelian Inheritance." *Transactions of the Royal Society of Edinburgh* 52, pp. 399–433.

Fisher, R. A. (1925). *Statistical Methods for Research Workers*. Oliver & Boyd, Edinburgh.

Fisher, R. A. (1935). *The Design of Experiments*. Hafner Publishing Company, New York.

Fisher, R. A., and Yates, F. (1963). *Statistical Tables for Biological, Agricultural, and Medical Research*. Oliver and Boyd, Edinburgh, England.

Fligner, M. A., and Killeen, T. J. (1976). "Distribution-Free Two-Sample Tests for Scale." *Journal of the American Statistical Association* 71, pp. 210–233.

Freund, R. J., and Wilson, W. J. (1997). *Statistical Methods*. Academic Press, San Diego, CA.

Giani, G., and Finner, H. (1991). "Some General Results on Least Favorable Parameter Configurations with Special Reference to Equivalence Testing and the Range Statistic." *Journal of Statistical Planning and Inference* 28, pp. 33–47.

Hochberg, Y., and Tamhane, A. C. (1987). *Multiple Comparison Procedures*. John Wiley & Sons, New York.

Hsu, J. C. (1996). *Multiple Comparisons: Theory and Methods*. Chapman & Hall, New York.

Kramer, C. Y. (1956). "Extensions of Multiple Range Tests to Group Means with Unequal Numbers of Replications." *Biometrics* 12, pp. 307–310.

LaPlace, P. S. (1827). "Mémoire sur le flux et reflux lunaire atmospheric." In *Connaissance des Temps pour l'an 1830*, pp. 3–18.

Layard, M. W. J. (1973). "Robust Large-Sample Tests for Homogeneity of Variance." *Journal of the American Statistical Association* 68, pp. 195–198.

Levene, H. (1960). "Robust Tests for Equality of Variances." In I. Olkin (ed.), *Contributions to Probability and Statistics*, Stanford University Press, Stanford, CA, pp. 278–292.

Marascuilo, L. A., and Levin, J. R. (1983). *Multivariate Statistics in the Social Sciences*. Brooks & Cole, Monterey, CA.

Mason, R. L., Gunst, R. F., and Hess, J. L. (1989). *Statistical Design and Analysis of Experiments with Applications to Engineering and Science*. John Wiley & Sons, New York.

Nelson, L. S. (1974). "Factors for the Analysis of Means." *Journal of Quality Technology* 6, pp. 175–181.

Nelson, P. R. (1982). "Exact Critical Points for the Analysis of Means." *Communications in Statistics: Theory and Methods* 11(6), pp. 699–709.

Nelson, P. R. (1983a). "A Comparison of Sample Sizes for the Analysis of Means and the Analysis of Variance." *Journal of Quality Technology* 15, pp. 33–39.

Nelson, P. R. (1983b). "The Analysis of Means for Balanced Experimental Designs." *Journal of Quality Technology* 15, pp. 45–54.

Nelson, P. R. (1985). "Power Curves for the Analysis of Means." *Technometrics* 27, pp. 65–73.

Nelson, P. R. (1988). "Testing for Interactions Using the Analysis of Means." *Technometrics* 30, pp. 53–61.

Nelson, P. R. (1989). "Multiple Comparisons of Means Using Simultaneous Confidence Intervals." *Journal of Quality Technology* 21, pp. 232–241.

Nelson, P. R. (1991). "Numerical Evaluation of Multivariate Normal Integrals with Correlations $\rho_{lj} = -\alpha_l \alpha_j$." *The Frontiers of Statistical Scientific Theory & Industrial Applications*, pp. 97–114.

Nelson, P. R. (1993). "Additional Uses for the Analysis of Means and Extended Tables of Critical Values." *Technometrics* 35, pp. 61–71.

Nelson, P. R., Coffin, M., and Copeland, K. A. F. (2003). *Introductory Statistics for Engineering Experimentation*. Academic Press, New York.

Nelson, P. R., Copeland, K. A. F., and Wludyka, P. S. (2005). "Robust ANOM Tests," University of North Florida Technical Report 050115. http://www.unf.edu/coas/math-stat/CRCS/CRTechRep.htm.

Nelson, P. R., and Dudewicz, E. J. (2002). "Exact Analysis of Means with Unequal Variances." *Technometrics* 44, pp. 152–160.

Ohta, H. (1981). "A Procedure for Pooling Data by the Analysis of Means." *Journal of Quality Technology* 13, pp. 115–119.

Ott, E. R. (1967). "Analysis of Means—A Graphical Procedure." *Industrial Quality Control* 24, pp. 101–109. Reprinted in *Journal of Quality Technology* 15 (1983), pp. 10–18.

Ott, E. R., Schilling, E. G., and Neubauer, D. V. (2000). *Process Quality Control: Troubleshooting and Interpretation of Data*. McGraw-Hill, New York.

Ramig, P. F. (1983). "Application of the Analysis of Means." *Journal of Quality Technology* 15, pp. 19–25.

Ryan, T. P. (2000). *Statistical Methods for Quality Improvement*, 2nd ed. John Wiley & Sons, New York.

SAS/QC® User's Guide Version 9 (2003). SAS Institute Inc., Cary, NC.

SAS/STAT® User's Guide Version 9 (2003). SAS Institute Inc., Cary, NC.

Scheffé, H. (1947). "The Relationship of Control Charts to the Analysis of Variance and Chi-Square Tests." *Journal of the American Statistical Association* 42, pp. 425–431.

Schilling, E. G. (1973). "A Systematic Approach to the Analysis of Means." *Journal of Quality Technology* 5, pp. 92–108, 147–159.

Sheesley, J. H. (1980). "Comparison of K Samples Involving Variables or Attribute Data Using the Analysis of Means." *Journal of Quality Technology* 12, pp. 47–52.

Sheesley, J. H. (1981). "Simplified Factors for Analysis of Means When the Standard Deviation is Estimated with the Range." *Journal of Quality Technology* 18, pp. 184–185.

Soong, W. C., and Hsu, J. C. (1997). "Using Complex Integration to Compute Multivariate Normal Probabilities." *Journal of Computational and Graphical Statistics* 6, pp. 397–425.

Stokes, M. E., Davis, C. S., and Koch, G. G. (2000). *Categorical Data Analysis Using the SAS System*. SAS Institute, Cary, NC.

Tukey, J. W. (1953). *The Problem of Multiple Comparisons*. Unpublished manuscript.

Vardeman, S. B. (1994). *Statistics for Engineering Problem Solving*. PWS Publishing Company, Boston.

Wheeler, D. J. (1995). *Advanced Topics in Statistical Process Control*. SPC Press, Knoxville, TN.

Wludyka, P. S. (1999). "Using SAS to Perform the Analysis of Means for Variances Test," University of North Florida Technical Report 990920. http://www.unf.edu/coas/math-stat/CRCS/CRTechRep.htm.

Wludyka, P. S., and Nelson, P. R. (1997a). "An Analysis-of-Means-Type Test for Variances from Normal Populations." *Technometrics* 39, pp. 274–285.

Wludyka, P. S., and Nelson, P. R. (1997b). "Analysis of Means Type Tests for Variances Using Jackknifing and Subsampling." *American Journal of Mathematical and Management Sciences* 17, pp. 31–60.

Wludyka, P., and Nelson, P. R. (1999). "Two Non-Parametric Analysis-of-Means Type Tests for Homogeneity of Variances." *Journal of Applied Statistics* 26, pp. 243–256.

Wludyka, P. S., Nelson, P. R., and Silva, P. R. (2001). "Power Curves for the Analysis of Means for Variances." *Journal of Quality Technology* 33, pp. 60–65.

Wludyka, P. S., and Sa, P. (2004). "Robust I-Sample Analysis of Means Type Randomization Tests for Variances for Unbalanced Designs." *Journal of Statistical Computation and Simulation* 74, pp. 701–726.

Wu, C. F. J., and Hamada, M. (2000) *Experiments: Planning, Analysis, and Parameter Design Optimization.* John Wiley & Sons, New York.

Index